Hugh Ridley is the author of books on varied aspects of German culture, law and history. After many years as a Professor at University College Dublin, he moved to the UK, where he now lives. He is a member of the Royal Irish Academy.

To the choughs at Rough Point, Maharees, Co. Kerry. That they should always flourish.

Hugh Ridley

BIRD PAINTING BETWEEN ART AND SCIENCE

The German Tradition

AUSTIN MACAULEY PUBLISHERS™
LONDON * CAMBRIDGE * NEW YORK * SHARJAH

Copyright © Hugh Ridley 2024

The right of Hugh Ridley to be identified as author of this work has been asserted by the author in accordance with sections 77 and 78 of the Copyright, Designs and Patents Act 1988.

All rights reserved. No part of this publication may be reproduced, stored in a retrieval system, or transmitted in any form or by any means, electronic, mechanical, photocopying, recording, or otherwise, without the prior permission of the publishers.

Any person who commits any unauthorised act in relation to this publication may be liable to criminal prosecution and civil claims for damages.

The story, experiences, and words are the author's alone.

A CIP catalogue record for this title is available from the British Library.

ISBN 9781398425934 (Paperback)
ISBN 9781398425941 (ePub e-book)

www.austinmacauley.co.uk

First Published 2024
Austin Macauley Publishers Ltd®
1 Canada Square
Canary Wharf
London
E14 5AA

I am grateful to the British Library for permission to reproduce plates 3, 11, 15, 16, 19, 20 and 21; to the Preußische Staatsbibliothek for permission to reproduce plates 6, 7, 8, 9, 10, 12, 13,14 and 22; to the Staatliche Kunstsammlungen Dresden for plate 2; to the Staatliches Kunstmuseum Schwerin for plates 4 and 5; and to the Kunsthalle Bielefeld for plate 24. Thanks too to the Albertina Vienna for plate 1. I am particularly grateful to the Konstmuseum Göterborg (and in particular to Eva Nylands) for plate 23, the Nolde-Stiftung for plate 24 and to Bernhard just of the wonderful Naumann Museum in Koethen for plate 17 and 18. The Naumann museum is a must for bird-lovers spending time in Germany and its helpfulness exemplary.

I would like to thank the wonderful staff of the rare books section of the Stabi, the BL and the Cambridge University Library. Their help and expertise hold up a model of what libraries are and should be.

I have plagued friends and family for many years with this project. For expertise and enthusiasm, Simon and Jenny, Sophie and Guy, who got it into print; Jochen Vogt, whose clear-headedness and generous interest have always inspired and encouraged me; the late Sebastian Neumeister, (the sad news of death reached my noly during the final stages of proof) whose wonderful book on Friedrich II was only part of the help he gave. Another eminent Hispanist, my friend, the late Don Cruickshank, helped me into emblems. My friend Richard Jones, a distinguished chemist, took time from the bridge table to help me with the periodic table. Jenny Uglow and Avril Pedley were another source of expert encouragement.

Finally, to the helpful staff of the AM production unit.

Table of Contents

Preface	11
Chapter One: Albrecht Dürer and the Little Owl	16
Chapter Two: Rembrandt: Self-Portrait with a Bittern	30
Chapter Three: Frisch: Smallest Little Owl (1763)	41
Chapter Four: Oudry: Painting the Menagerie	55
Chapter Five: The Last Great Natural Historian: Buffon's Natural History and Its Impact on German Ornithology	67
Chapter Six: A Dangerous Obsession: Nineteenth-Century Classification	92
Chapter Seven: A Peasant Farmer Goes to School	108
Chapter Eight: Josef Wolf Takes London by Storm	124
Chapter Nine: Watch the Dickie Bird: The Bird Picture in the Age of Photography	139
Chapter Ten: On the Edge of Modernism: Liljefors and *Nolde	150
Notes	169
Brief Biographical Notes	193

Preface

A presentation of German bird painting to an English-speaking readership is long overdue. For centuries, English bird watching felt itself to be the world centre of the activity, a status increased by Britain's extensive overseas possessions and the species found in these territories. In addition, the closeness of bird watching to ornithology as a science led to an assumption of a general English language pre-eminence in the science too, while London was such a focus of bird painting talent in the nineteenth century that other countries were seen as peripheral. In any case, English language interest in Germany was sporadic, focusing on things such as chemistry and religious studies (later with a bit of physics and fascism thrown in) – none of which did anything to make German ornithology better known.

In retrospect, it appears foolish to speak of nationalities when the language of painting seems indifferent to everything, save to form and colour and when birds are citizens of no country but this book is not about identities or nationality. I include a chapter on Josef Wolf, the German painter, who took London by storm in the mid-nineteenth century and became in many respects the spirit of English bird painting but I am not fighting for his soul or examining his passport. His career does imply a merging of national traditions and underlines the internationalism which properly is at the heart of the subject.

The challenge of this book is to show as fascinating and worthwhile the development of bird painting in Germany across 400 years. I present significant paintings, representing major figures, discuss the science background from which they come and suggest some artistic issues which they raise. Above all, I hope readers will be fascinated by the pictures themselves. Whether the birds lurk in the margins of medieval manuscripts or adorn the full size plates of modern hand books, the self-sufficient beauty of the various species affords equal delight. The power and energy of Killian *Mullarney's merlin or Lars *Jonsson's eiders and curlew look different from what we find in manuscripts or in *Fabritius' goldfinch but they share the same fascination. Even within a short book, I wish to emphasize the richness of the field – the huge numbers of bird books, let alone bird illustrations published over the last 500 years. There are some extraordinary pictures to be explored, some fascinating debates and some fine achievements to be noted. This book is only a start.

Many names discussed here have faded from English language texts but they were once important, not least to my predecessors in the field. Not just Jean Anker in her bibliography but Christine E. Jackson and Maureen Lambourne have found the history of bird painting meaningful only with the inclusion of French, German, Italian and American names. The scientific field too has kept its internationalism but the popular aspect has been not maintained. My own account wishes to remain free of any cultural nationalism and partly for that reason (but principally for their huge interest and wonderful pictures), I include in my narrative a Dutch and a Swedish painter and though I concentrate on the Germans, half-a-dozen French artists form part of my narrative. No worthwhile history can operate without open frontiers.

Over recent years, three books have begun to suggest that the imbalance may be correcting. In *Birds in a Cage* (2012), Derek Niemann told the story of four British prisoners of war, discovering in captivity in Germany the presence there of a strong ornithological community represented by 'the most influential ornithologist of the twentieth century', Erwin *Stresemann. Stresemann came into their story not simply as a courageous and humanitarian figure but as the focal point of centuries of German ornithological history. It should be interesting to learn more about the tradition from which he came. In 2015, Angela Wulf's major biography of Alexander von Humboldt showed the world importance of that great scientist and explorer and the richness of the scientific tradition from which he came. It was striking that even a massive figure, such as Humboldt, could be seen in the English speaking world as a 'lost hero of science'. This welcome focus was reinforced when the distinguished British ornithological scientist, Tim Birkhead, in a study of the early ornithologist Francis *Willughby, turned his attention to the German background. It had always been known that, together with his colleague John *Ray, Willughby had acquired the manuscript of Leonhard *Baldner's inventory of the birds he had seen during his work as a boatman on the Rhine near Strasbourg (dated 1666). The self-same manuscript has long been in the British Library. Birkhead subjected Baldner's work to full scholarly analysis and the relationship between three pioneering ornithologists across frontiers and languages received its first appropriate treatment.

The first two chapters discuss masterpieces of bird painting by Albrecht Dürer and Rembrandt van Ryn. These studies convey a sense of where bird painting began and the obstacles which it overcame. Chapter Three introduces a pioneering bird illustrator and the handbook in which his work appeared in the mid-eighteenth century. The illustrator, an engraver in Berlin, and the lead author of the handbook, Johann Leonhard Frisch, produced a remarkable piece of science and fascinating pictures. While Rembrandt's picture was painted during the religious wars which afflicted the continent from 1618–1648, Frisch's book appeared at the end of another long war (this time Seven Years) which had shaken his country. British prisoners of war were not the only people to discover that bird watching can help one over the calamities of history.

Moving forward in time, Chapter Four introduces a painter who represents in Germany the heights of courtly painting as practised in Versailles. His work shows how a primarily decorative painting style can embrace, albeit in a rudimentary form, some of the scientific developments of its day. The French influence continues to be a topic in the fifth chapter. The monumental natural history produced by the Comte de Buffon (seven volumes alone on birds) was a major event in the scientific and philosophical life of various countries, including England and Scotland. It epitomises an enlightenment approach to nature and we consider a range of the illustrations which adorned it in its German translations after 1770.

Among the developments affecting ornithology both as the academic science it had become in Europe and as a field activity, classification was the most dominant. Chapter six looks – not too seriously, for some of the ideas were, to say the least, freakish – at what Harriet Ritvo called 'Figments of the Classifying Imagination', focusing on the influence of the German zoologist, Lorenz Oken, on English ornithology and showing how classification affected bird painting at the time. This line of influence parallels the huge popularity of German Romantic philosophy among English poets and intellectuals at the beginning of the nineteenth century. To remind us of the scientific importance of the work of the classifying ornithologists, the chapter contains a comparison with the ground-breaking work of chemists in classifying and systematising the chemical elements in the periodic table.

Chapters Seven and Eight focus on the achievements of two sharply contrasted but remarkable bird painters of the nineteenth century: Johann Friedrich Naumann and Josef Wolf. These chapters offer a view of the contrasting circumstances under which bird painting and ornithology itself could flourish in England and Germany, as well as highlighting the contemporary debates on bird painting caught between science and art. In addition, Chapter Seven places ornithology – here from its formal side as writing – in a wider scientific and literary context.

From the mid-century, photography was the elephant in the room for the bird painters. Artists used photography, increasingly as it became less cumbersome but were scared of it. Its champions made claims for its future importance in the representation of the natural world. Chapter Nine examines these claims in the context of bird photography.

The final chapter moves into an art world far from the photographed natural world. Its first focus is a remarkable picture by the Swedish artist Bruno Liljefors, responding both to his training as a nature painter in Germany and to the inspiration of Darwin, mediated through a Danish writer and naturalist. The suggestion is that what may appear to be the least scientific of the pictures so far (for it lacks 'photographic realism') is steeped in science. The final picture painted by Ernst Nolde in the South Seas on the eve of the First World War brings the discussion back to the symbolism and exoticism of some of the earliest birds to feature in European painting. Nolde did not change the history of bird painting. His work in no way marks an end to traditions of bird painting proudly continued in the intervening century – but it rounds off developments in the nineteenth century and acts as a reminder of the perennial charm and fascination of bird painting.

A book on bird painting will have many possible starting points, from chance encounters in a gallery to sightings on wind-swept headlands. Everywhere and always the beauty and variety of bird species break into our lives and we wish to hold on to that beauty rather than watch it take flight. In Ireland, bird painting is made real by the power of its master, Killian Mullarney, but also by the younger generations of bird watchers as they turn the sketches brought back from field-trips into more substantial pictures. For bird watchers, painting is a very real activity. I have never forgotten listening to my son's friends comparing their pictures and discussing why my son had painted ivy leaves in the foreground of his picture while a friend had preferred barbed wire. These pictures begin in science but end in aesthetics. This, for 12-year-olds, was the sharp end of a centuries-long debate on the relation between art and science in bird painting.

As we follow that debate back into history, its two terms are subject to considerable change. Ornithology developed from a pastime of country folk into a rigorous scientific discipline. It moved in and out of the laboratory as it developed. The history of ornithology covers key areas of zoology and the life sciences in general but has never lost that mixture of field observation and laboratory science which has left room for the amateur and enthusiast. Art – its styles, its purpose and its place in society – has hardly remained unchanged either. Our debate takes place on a shifting ground.

For whatever reason, each generation has been entranced with bird portraits. As I worked on this book, I became ever more conscious of how many bird painters of past centuries remain unknown today and not just in Germany. There are so many names which would reward rediscovery. Few pleasures are more intense than to re-discover one of these collections, to open a book of coloured plates – often, it seems, unseen for hundreds of years – and experience the freshness and life enclosed for so long in its pages. Bird watchers have their rarities and their ticks but so does the historian. But this book is not aiming at completeness and at best tries to draw attention to (rather than exhaust) the richness of the field.

At the end of the book I include brief details on the names unfamiliar to readers (an asterisk before a name in the text at its first appearance indicates that biographical information is available but the information is rudimentary and if readers recognise a name they should not bother to follow it up). I present my text without footnotes. German matters are largely unfamiliar to English speaking readers and I have erred on the side of explaining the obvious. All sources are given in the end notes – including German language material and giving English sources wherever practical – but this book is about a place where scientific history overlaps with art and readers will get more from looking at the pictures than from straining their eyes over footnotes. The end notes to this preface give the bibliographical background to the whole topic and those readers who wish to explore further can start there.

Chapter One
Albrecht Dürer and the Little Owl

A complete history of bird painting would have to start in pre-history in the caves of Lascaux and Chauvet or later with the sacred birds of ancient Egypt. This book has a smaller period in mind and a narrower focus. Its subject is a broadly defined German tradition, starting within an artistic movement and then following the relationship between art and science as it emerges in the work of a range of ornithologists and artists between 1500 and 1900.

Bird pictures were not just a product of human wonder at the beauty of birds but they were part of a still deeper feeling: the wish to understand nature. The subject thus involves the relationship between the sense of wonder expressed in art and the shifting state of knowledge about nature – natural history as it was initially called, science as it came to be known – and about birds in particular. Our period reaches from the early modern through the Enlightenment and into the scientific revolution of the nineteenth century. Over these years, styles of art changed considerably as a succession of artists brought birds on to their canvas, while ornithology transformed itself from a hobby of the leisure class into a sophisticated scientific discipline.

We cannot examine this relationship between art and science without including a further dimension: the history of scientific illustration. For as science developed in the nineteenth century, acting as the driver for a massive popularisation of knowledge and information, illustrations played a central role, both in analysing and exhibiting the objects of study and in the general spread of science. It's well-known how important pictures were in the dissemination of new ideas. This was no less true in science than it had been in spreading religious ideas in the preceding centuries and we need to consider the relationship of art to scientific illustration more critically.

In a standard work on the history of book illustration, Claus Nissen points to the fluid boundary line between scientific illustrations – often works of extraordinary skill and beauty – and traditionally defined art. In this book, I do not see the domains as opposites, on the contrary, in bird painting they seem to merge. At what level can one really differentiate between a beautifully executed illustration and a work ripe for the gallery? And if so, in what way: a picture's function, for instance, or its craftsmanship, its 'beauty', or – more crudely – its subject? I will suggest ways in which the scientific intention behind a picture intensifies its expressive power; at other times, it is the painter's artistic principles which create the scientific value of a picture. The bird artists of the present celebrate in their work the harmony of these human drives and in doing so they continue the achievement of past centuries.

Illustration contains at least two dimensions. For the bibliophile, it is a decorative art of great beauty while, for the scientist, much illustration has nothing to do with beauty, simply with recording knowledge. Bird painting will tend to combine both of these dimensions and it reminds us that, at its most basic, illustration has always understood itself as a way to understanding. Drawing – a skill routinely learnt in previous centuries – was always seen as the key to the natural world to understanding how plants and animals worked, their essence. *Goethe recommended a friend to take up drawing 'because it will mean that for the first time you become conscious of nature'. The other great poet of the German eighteenth century, Friedrich *Schiller, trained in medicine and history, also spoke of the importance of 'aesthetic education'. Jean-Jacques *Rousseau too believed that 'our first teachers of philosophy are our feet or hands and our eyes'. Horst Bredekamp summarised this in his phrase 'thinking hands'. I would like to show how in bird painting objective knowledge, aesthetic sensibility and draughtsmanship combine, informing and delighting.

It was not easy for bird painters to keep up with the science. Since the eighteenth century, ornithology has been in the avant-garde of scientific biology. Many of important paradigmatic shifts in biology have started out there – it is truly a pioneer science. This status has many causes relating to fundamental features of birds – the number and variety of species, their complex social forms and the breadth of their adaptions – but it is also a feature of their interrelation with humans. Humans and birds live close to one another and – though birds seem reasonably uninterested in humans – fascination with birds is where the science starts. The basic questions raised by bird studies form an impressive list, some of which impinged on the work of the painters. These advances include: Darwin's theories on the origin of species or on sexual selection, issues of instinct and intelligence in animals, which were directly raised in considering bird migration, population studies which have variously focused on birds (the enigma of the number of eggs which birds lay and its relationship to particular living conditions from year to year), the strategies of camouflage and survival – it's not just the canary in the mine which has been used to monitor the world we live in. Ornithology was always exemplary in the directions in which it took science and – no less importantly – its achievements almost invariably reached across the broadest spectrum of contributions from high science to fascinated hobby observation.

Without the physical and emotional closeness of humans to birds, the relationship between professional science and the amateurs would not have been so close. Although from the mid-nineteenth century, there has been an occasional tension between field work and the laboratory, ornithology has never ceased to depend on the observation of birds in nature – an activity in which specialised scientific interests have harmonised smoothly with those of hobby bird watchers. If anything, the growth of ethology (the science of animal behaviour) as a discipline from the late nineteenth century on has made that harmony still greater, especially when combined with the science of ecology, which has increasingly appealed to a broad public.

It has been the closeness of humans to birds which has meant that from the sixteenth century on, many scientific publications – for instance, strikingly, the great work of the French naturalist *Buffon – were addressed to a double audience, professional and amateur. The amateurs were not just potential purchasers of the books – they were involved in supplying the observations on which the books were based. Pictures were one way in which this mediation between professional and amateur took place. So, some of the images we discuss in the chapters of this book seem deceptively clear and transparent. Increasingly, however, they occupy an ambivalent position between art and science, anthropology and zoology, empathy and distance. This ambivalence is a recurring theme of the book.

There is a long history behind discussions of the connection between science and aesthetics. It was not just Bible readers who believed that understanding nature would reveal the beauty of the Creator and who therefore looked to harmonise science with art. Beauty and understanding belong closely in other sciences too, for instance in mathematics, as anyone knows who has heard a mathematician praising the elegance of an equation – usually one which the lay person does not follow. That relationship was part of the founding principles of the Enlightenment. In one of the cult books of the early nineteenth century, the world-famous explorer and naturalist, Alexander von *Humboldt – whose diaries accompanied (perhaps even inspired) Darwin's journey on the *Beagle* – argued that to understand nature, even to classify it, required an artist's eye, only such an eye could bring order to what it sees in nature. The painter, by his work, identifies the 'specific physiognomy' of a landscape (by 'physiognomy' Humboldt understood precisely those features which the science of ecology was to reveal, notably the link between forests and water-flow or the nature of desertification). A painting which depicts a plant possesses a similar skill. 'The artist has the gift of differentiating groups,' Humboldt writes. 'Under the artist's hand, the magical complexity of nature […] is resolved into a few simple features.' He thus equates the expressive ability of art with the ability to perceive objective scientific categories; in other words, with fundamental taxonomy. Humboldt shared with his close friend, Goethe, an intense feeling for the characteristic morphology of plants – a feeling which inspired Goethe, as poet no less than as scientist. To learn to see could be both artistic and scientific. As science became more specialised in the course of the nineteenth century, it became difficult to sustain Humboldt's vision in its original form. It's striking to read of the Danish biologist and novelist, J.P. *Jacobsen, that he possessed 'the ability to look at a flower or plant from the perspective of an artist and scientist simultaneously'. Not all bird painters could do that and only the best aspired to.

To see art as a route of analysis is not a common approach to bird painting (or to nature painting in general) and it takes this book on a route which others avoid, for instance via seemingly un-aesthetic topics such as taxonomy or Darwinism. When we include scientific understanding as part of bird painting, we have also to take in less successful pictures which show the relationship between art and science as tension rather than as fulfilment. The historical progression of the chapters thus reflects a learning process on behalf of the artists, a process which is as far from finished in the present time as is the fascination with birds itself.

In a stimulating essay examining the relationship between scientific illustrations and the scientific principles of their age, the art historian, Martin Kemp, writes: 'In order to read these pictures properly, we must learn to understand what demands the illustrator is making on us. He pleads for our understanding as he tries to demonstrate his knowledge and he trusts us to approach his work with the appropriate interpretative tools.' I cannot claim to live up fully to Kemp's exhortation but I hope to use his insight to unlock the door to some of the bird pictures which clamour for our attention.

The Familiar and the Other: Understanding Nature in the Early Renaissance

Even at the beginning of the early modern period, ornithology already had its own history. This had developed from the writings of the ancients, above all, *Aristotle and *Pliny, authorities whose importance was to be enhanced by the Renaissance. The Bible supplied ornithology with a few stories and rather less science but it made up for its gaps by offering a multitude of images from which early bird painting could develop. The stories of the creation and the paradise garden inspired artists in many ways, justifying their sense of the specialness of nature. With the story of the flood, artists were encouraged to set about creating their own inventory of animals and birds, turning their art into an Ark of the imagination. We can see that shift as late as the mid-nineteenth century in the work of Edward *Hicks, for whom 'the peaceful creation' was a topos repeatedly productive of animal and bird pictures. Animals and birds decorate the buildings and manuscripts of the middle-ages, only then to become a central accompaniment of European painting in the Renaissance.

The Bible stories offered a combination of exotic and familiar elements, out of which European art increasingly preferred and encouraged the familiar. Nativity scenes become portraits of domestic life in the homelands of the artists, Galilean shepherds looked like Tuscans and Saxons. It was not until the Europeans set out to conquer the world that the exotic plants and fauna of a non-European environment began to appear on their canvases. The complex relationship between the exotic and the familiar was played out for the following 400 years in the minds of ornithologists and it raised specific questions. Were European birds the highpoint of world avifauna? On that basis, it was argued at one time that North American birds were inferior to their European fellows, indeed degenerate. Buffon, for instance, argued that migration 'produced important alterations on their (i.e., birds) external appearance and perverts their instincts' – an opinion which drove *Audubon to his magnificent artistic project. More narrowly, should European ornithology turn its eyes from their own hedgerows and skies?

German ornithology – in its early days – followed a domestic model more closely perhaps than was the case in other countries. Significantly, unlike Britain or the Netherlands, Germany had no overseas territories and little hope of acquiring any. The natural historians and painters concentrated above all on their home patch, the upper Rhine in the case of Leonhard *Baldner or the Saxon castle and its park for Johann Friedrich *Naumann. For Baldner, even birds of the Baltic were rarities. In the seventeenth century he had not the slightest idea of what might be found beyond his narrow horizon. But he made the most of his situation and, by concentrating his work on the birds he had actually seen, he established a tradition to counter that of the early animal books – the so-called Bestaria – which had no interest even in differentiating between real and legendary creatures. Baldner's world was too small for the exotic, yet it was big enough for inspiring bird painting. A hundred years later the painter Naumann took a route which was even more inspiring. As a peasant farmer tied to his poor land, he could neither travel far, nor widen his horizon by any method other than that of science to which he assiduously devoted himself with the help of a nearby university. Only when he became famous could Naumann travel beyond his immediate environs – even then not leaving Europe – but, like many Germans, when he did travel, he took his science with him.

For such reasons, German ornithologists had little chance to go beyond the inventorying of their local birds. It wasn't just a question of personal circumstances, their situation was determined by the fragmentation of Germany and its lack of any distant territories. As late as 1779, when in England scientists were starting to realise the extraordinary classification tasks posed by Australian mammals and monotremes, the holder of a prestigious Göttingen professorship, Johann *Blumenbach, identified the most urgent task of biologists in his generation to establish what he called a fauna Germanica.

For all these reasons, I feel confident in beginning my account neither in Egypt nor in a French cave but in Nuremberg in 1508 with Albrecht Dürer and his *Little Owl* **(Plate 1)**. His work is one of the best known and most loved bird paintings in European art. We can use it as the entry to our topic.

Dürer's watercolour portrait: *Little Owl* – a first consideration

I should perhaps make clear at the start that Dürer experts do not regard this picture as one of his best. If one compares it to his famous *Hare* (1502), differences in brush work and observation of detail become clear. The eyes too are disappointing – perhaps the colouring is wrong, perhaps they are a little lacking in life. There's also a significant difference between this picture and the sensual delight with which Dürer painted the dazzling colours of the blue roller (*Dead Blue Roller*, watercolour, 1512). The little owl is rather every day and unassuming. There are even critics who doubt that Dürer painted it himself. It does nothing to our argument if he should not have been the painter, for the subject is more important than the artist. For the discussion, however, we'll stay with the assumption that Dürer did paint it.

Plate 1: Dürer, *Little Owl*

The picture is important because it is a portrait. We will use the word 'portrait' throughout the book and justify it by pointing out that modern art historians use the term to denote the flower pictures of the Dutch masters. A picture which concentrates on one bloom (for instance, during the tulip mania of the early seventeenth century) deserves a more precise name than simply 'picture'. 'Portrait' makes explicit the individual worth of the subject, a sense of worth shared between painter and public, that is certainly the case with the bird pictures discussed in this book. In her important study of Darwin's relationship to his painters, Julia Voss finally concedes to the work of Edward *Landseer and Josef *Wolf the title of portraits and sees in the first half of the nineteenth century a breakthrough in painting. 'Animals became worthy of portraits,' Voss writes. Dürer's *Little Owl* makes clear that we don't have to wait that long to find portraits. The breakthrough came 300 years earlier.

There's another reason why portrait is such an appropriate term. The birds have a much longer ancestral title than the ephemeral statesmen, landed gentry and titled young women who adorn so many canvasses. '40 centuries look down on you,' Napoleon told his soldiers as they entered Egypt. When Cleopatra sat on her throne – like the Shah on his thousand-year peacock throne – the birds had been there ten times longer. If anyone deserves a portrait, then it is the birds.

Bird painting before Dürer

There are numerous birds painted in the manuscripts of the Middle Ages and in the work of the early Italian Renaissance. Some of these are painted so accurately that their species can be identified. If we look through these pictures – perhaps using Brunsdon Yapps' beautiful catalogue – it's at once clear that these birds are not portrayed for their own sake. They either accompany a Bible story, offer a decorative fill-in for an empty corner or take over a wider symbolic function. So it is that an eagle invariably watches over St John's Gospel, adorning title pages and decorated initials, even though no eagle is actually mentioned in that gospel. Other biblical stories are more specific in their demands for bird painting, such as the scurrilous story of Tobias (in the *Apocrypha*, but no less well-known for that). Tobias (or Tobit) was disobedient to God, who then punished him with blindness, using as the instrument of his punishment a swallow's droppings into his eye. Illustrators had no difficulty in reproducing the details of this action (after all, scatology was a major part of the religious polemics of the Reformation and excretion an activity much favoured in portraying one's enemies) but the story challenged them to paint a swallow and we can indeed recognise swallows in the many representations of the story but the illustrators are hardly interested in the swallows. The particular species arrives in their pictures – to misquote Brecht – as accidentally as Pilate arrived in the Creed.

A particular favourite of the illustrators was a scene from the *Book of Revelation*. As the Apocalypse gets under way, an angel calls out with a loud voice to the birds of the air and instructs them to eat the corpses of the executed kings. We should apparently speak not only of the Knights of the Apocalypse but also of its birds. But even then, these birds do not matter in themselves – unlike in a modern apocalypse, where the birds would certainly wish to gorge themselves on the humans who had done so much to wipe them out and destroy their lives. The Bible story is more about dead kings than live birds.

Only one source before Dürer escapes this indifference to the birds themselves. It dates from the early thirteenth century when the Holy Roman Emperor, the German *Friedrich II, for some reason resident as King of Sicily, found the time to compile an extraordinarily significant list of the birds he had observed. Scholars have used his works not simply (as is usual) to chronicle the failures of early bird identification but on the basis of the evidence Friedrich provides (and through an exciting new technique of historical ornithology) to point to subsequent shifts in the species of birds inhabiting the North-East Mediterranean at and since the time of Friedrich. Rather like the distant emperor, Marcus zum *Lamm, not overwhelmed by his clerical and legal duties in the modest town of Heidelberg, prepared a significant manuscript on the birds he had observed in his area. When birds are real, bird watchers can find that their religious and professional duties take second place.

Dürer's picture predates Lamm by some 50 years and he had no knowledge of Friedrich II's labours. (These were brought to light at the end of the eighteenth century by Blasius *Merrem – whose work we encounter in a later chapter) but his *Little Owl* shares the attitude of these two figures. For Dürer, the little Owl needs no human or biblical reason to stand for his portrait. He is worth painting for his own sake – his ordinariness is a recommendation. Dürer could not make it more evident here that he is not interested in exoticism. Nature's fascination is in its simple existence, its absence of transcendence, it doesn't need to be symbolic to be important. In taking up this position, Dürer anticipated much of the so-called realism of the nineteenth century. When he writes, 'The closer you can get to representing nature and life, the better and more artistic your work will be', we could imagine we were reading Naturalists, like *Zola or in Germany Arno *Holz. Even in 1508, true to life representation and art were in harmony.

Dürer's various animal portraits seem to have been prepared in the same spirit. This is especially true of his many pictures of dogs – clearly an animal he greatly loved – from the Emperor Maximilian's hunting hounds down to the street dogs of Nuremberg. He was interested in exotic animals – as witnessed by his famous etching of the rhinoceros – but everyday animals are in a huge majority. Connoisseurs may find the little owl short on delicate brushwork and feel that its colours and details fail to come up to the roller but Dürer's bird has one huge advantage over that remarkable picture: it is alive.

We shall repeatedly return to this point. The blue roller is painted as a corpse, indeed as part of a skin. It's not a hunting trophy (though Dürer did paint hunting pictures), just a cadaver. But the little owl is painted as a living bird, we may suspect that in fact it was painted from a skin but the portrait aims to bring it back to life. In 1848 the celebrated ornithologist Hermann *Schlegel wrote a treatise on the art of bird painting (more on his ideas in Chapter Seven) and among other things drew attention to the difficulty of painting birds' eyes convincingly as 'living'. Dürer's portrait does not quite come up to Schlegel's demanding standards in this respect. We shall see that elsewhere the description of a bird painting as being 'according to life' did not distinguish between living and dead models. Yet Dürer's portrait shows an interest in life for its own sake and that has to be a criterion for all bird painting.

Symbolism or: A bird is a bird

This picture has a second feature which makes it an essential introduction to this book. It manifestly has no other subject than the bird itself. It isn't decorative or the illustration of some historical or religious constellation. It may lack its natural habitat (painters were in any case slow to concern themselves with this feature of their subjects and in that respect they reflected some of the scientists' priorities) but it occupies its own space. In contrast to the Rembrandt picture, which we examine in the next chapter, Dürer's picture has no connection with humans. If the owl is in captivity, for instance, that's not referred to.

The absence of symbolism presents a more serious problem; indeed, this feature of the picture must have puzzled its first public in 1508. In the Renaissance owls enjoyed a high level of symbolic value. From classical antiquity, the owl, as the bird of Minerva, represented intelligence and wisdom. Despite all recent research which ascribes to the owl a level of intelligence considerably inferior to that of other birds, the owl has remained a kind of feathered patron saint for teachers, scholars and exam candidates but Dürer is not interested in that.

Christian symbolism was even more specific about the owl than classical antiquity had been. Many contemporary pictures showed the owl in daylight being mobbed by other birds. This reference to suffering and unjustified torment was then focused on the figure of Christ. Dürer used this motif in other pictures (it had been a favourite theme of his revered teacher, Meister *Schongauer) but it is conspicuously absent from this picture. Whether or not the owl is in captivity, it is certainly not suffering.

What's remarkable about this freedom from added significance is the fact that Dürer was working in an age otherwise dominated by religious symbolism. Contemporary artists – and no less compulsively, art historians examining the period – are challenged by the conviction that, if nature proceeds from the hand of God, then every picture of nature must, in order to be truthful, convey this second level of meaning. God's signature is in all nature, indeed in everything, and must be shown as such. Art historians have therefore to assume, for instance, that every time the colour red appears in a bird's feathers, it is a reminder of (and therefore a reference to) Christ's crucifixion. You can accept this more readily in medieval pictures, where colour is doled out to the birds in a far from naturalistic way and where not all species can be identified. But, from the artist's point of view, if a colour is in itself symbolic, then it is all but impossible to paint anything which does not have symbolic importance – a situation which plays havoc with any sensible bird painting. Indeed, one art-historian, frustrated by this situation (how can one deal with anything when it is really something else?) concluded by a process of elimination that there was only one bird which was 'without symbolic significance': the siskin. Such was the world view surrounding Dürer and in which he had to work and a hard situation for someone who doesn't want to paint symbols, but in his *Little Owl,* he effortlessly leaves that world behind. The bird is not red because it is not red; it doesn't suffer because it doesn't suffer.

There's one more form of symbolic intensification of pictures to which we turn more fully in a later chapter. It involves the ability of art to represent and celebrate power and it is as accessible to the bird artist as to anyone else. It is the subject of a commentary by Simon Schama on a celebrated wood-cut by the bird artist, Thomas *Bewick. Towards the end of the eighteenth century, Bewick had created an impressive wood-cut showing a magnificent bull. Schama sees in this picture 'the great, perhaps the greatest icon of British natural history, and one laden with moral, national and historical sentiment as well as a purely zoological fascination'. A mixture of power, pedigree and pride capable of representing the whole nation therefore. England was not the only nation sensitive to such symbolic enhancement and as we read a rather patriotic account of Dürer's plant and animal paintings, we find a similar view. The author bemoans the fact that Dürer never painted a 'proper' picture of an eagle, this 'the heraldic animal of the German Reich, a creature with which none other can compete in size or majesty, powerful flight and importance'. With his *Little Owl,* Dürer resisted not just religious and classical symbolism but took care to make no concessions to those who would try to lay claim to his work in other causes. It all amounts to a good reason to be grateful to the little bird.

How then did Dürer manage to paint something so defiantly un-symbolic and so naturalistic?

Dürer's View of Art

Research has tended to the view that Dürer's style – we might call it documentary naturalism – goes back to a visit which he paid to Italy in 1490, during which he familiarised himself with the painting of the Italian Renaissance. When he came back to Nuremberg, his first commission was to make copies of Leonardo's horse drawings. It's possible that Dürer had also encountered Leonardo's anatomical drawings, which have inspired scientific illustrators throughout the centuries. Another important contact was with the painter Jacopo di *Barbari, renowned among other things for his bird pictures. Dürer's greatest gain from his Italian journey was the conviction – to quote the anatomical painter of the early nineteenth century, Charles *Bell, that the eye rather than any text 'spoke the real language of the object'. Dürer's love of nature pictures blossomed during these years. Important additional impulses came from the early Dutch masters, notably Jan van *Eyck but also from Schongauer and from the unknown artist hidden behind the title *Master of the playing cards, among whose copperplates there are some successful bird pictures.

Perhaps the greatest inspiration for Dürer's bird studies came from the contemporary fashion for plant books. In contrast to the Dutch flower pictures of the seventeenth century, the plant books were no luxury items but practical handbooks describing plants which had a medical use. They were of considerable economic significance, for they helped apothecaries and doctors accurately to identify the plants from which herbal remedies were extracted. Dürer's own plant pictures seldom have any medical relevance but they do show his mastery in a genre which demanded great accuracy of detail and which was entirely practical in intent. Birds did not achieve a similar level of practicality – only falconry and gastronomy offered financial rewards similar to those in medicine – and for that reason it was about 200 years before ornithology books began to catch up with the progress made in botany. Dürer had taken the scientific spirit of botanical illustration and applied it to another subject. We might compare the roller wing (1512) with Leonardo's detailed anatomical studies; in both areas (as in botanical painting) there was a harmonious relationship between draughtsmanship, art and science.

We return finally to the *Little Owl*, a bird Dürer has shown worth of its portrait with no need for the false feathers of religious or classical symbolism. Dürer's successor, the painter Lazarus Röting, all but forgotten today, continued his bird work while another pupil, Hans Weiditz, stayed in botanical painting. But in the long run it was the great scientific illustrators of later centuries who profited from Dürer's example and built on his technical achievements and on those of the whole Nuremberg school (this was especially noticeable in the treatment of perspective). But Dürer's other successors were all the painters and drawers who saw nature as a worthy object of art. All of them are inspired, directly or indirectly, by Dürer's motto: 'Truly art is in nature and whoever can find it there, nature will never let go'.

After this beginning, neither science nor art developed along a straight line on the example of Dürer's naturalism. Symbolism's hold on nature hardly slackened and the sciences had a long struggle before they could free themselves from religion. Other value systems intruded on nature painting while art kept redefining itself in ways which did not involve the imitation of nature. But anyone who followed Dürer's example would not forget his inspiration. Dürer's

Little Owl has helped hatch out seemingly endless generations of new bird painters.

Chapter Two
Rembrandt: Self-Portrait with a Bittern

We turn now to a celebrated picture from 1639 (**Plate 2**), a relatively early period of Rembrandt's career. It is one of 75 self-portraits which he is known to have painted. The eye is immediately seized, however, by the luminous colours of the bittern, allied to the care with which the bird is painted. While Rembrandt's characteristic dark shades are in the picture, the bird lover rightly feels the wonderful textures and the nuances of colour in the feathers to be at the centre of the picture. The bittern is certainly not an appendage to the human portrait but is the focus of Rembrandt's presentation. The human figure remains enigmatically in the background. To explore this picture, we encounter some of the ambiguities of bird painting in the Renaissance and will come to see interesting differences and similarities to Dürer's *Little Owl*. Unfortunately, the most obvious contrast is that the bittern is dead.

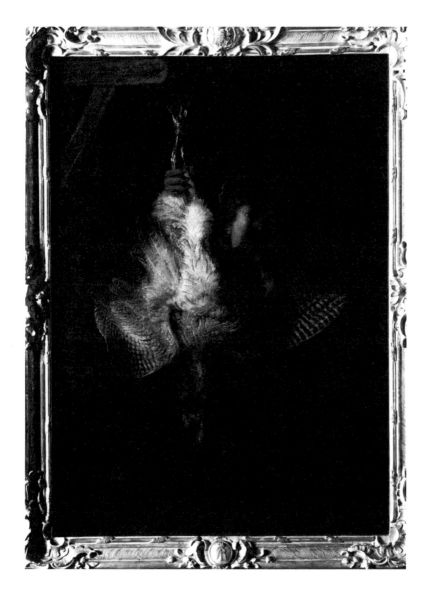

Plate 2: Rembrandt, *Self-portrait with a bittern*

Symbolism

Had this picture been painted a hundred years earlier, interpretations of the picture would have focused unambiguously on the various forms of symbolism associated with the species in whose company Rembrandt chose to portray himself. The bittern is indeed one of the species specifically, if infrequently, mentioned in the Bible, where the references seem to emphasize the bird as a harbinger or expression of desolation, a bird close to the night. Indeed, Isaiah does not seem to differentiate the bittern clearly from a more common night bird, the owl. (Such are the bird name problems across languages – let alone ancient languages – that all translations will tend to be no more than approximations as Buffon was to argue.) When Isaiah describes judgement coming down as the Lord delivers 'all the nations [...] to the slaughter', he prophesies that the land will become a waste land, a land possessed by 'the pelican and the porcupine [...] the bittern and the raven'. Perhaps it is the bittern's secretive disposition and its preference for a lonely marsh land habitat which caused its associations with disaster; perhaps too the bittern's boom seemed an avian form of the trumpets of judgement, a negative counter to the larks' voices, which were destined to join in the melodies of the heavenly choirs.

Symbolism of this kind, based on an automated reading of biblical and classical sources, in which the bird is immediately understood within a specific moral frame, had declined somewhat by the time Rembrandt painted. Despite this decline, however, the traditions of the emblem books had not been forgotten. Emblems consisted of a picture and a short text, interpreting the object or action in the picture and suggesting a moral. Many emblems were based on birds (usually primitively drawn) but their influence on zoological textbooks, including *Belon and *Gesner, was explicit. Consequently, the modern reader cannot afford entirely to ignore their possible relevance when looking through the history of bird paintings – no more, incidentally, than the student of the modern lyric can afford to ignore the role of traditional flower symbolism. Even starkly modernist poems evoke that ancient symbolism, just as elements of it remain enshrined in present day commercial flower arranging, with their conventions and fixed associations. Back in the seventeenth century such symbolism was easily revived after a period of lapse, as was shown when the genre of the so-called *Vanitas* portrait with birds was reactivated some years after Rembrandt's self-portrait. The dead bittern might have resumed its function as a powerful reminder of a coming judgement, a *memento mori*. At this time the depiction of the simple duck – in part because its Dutch name, *ende*, in itself recalled finality – often carried this hidden significance too.

Despite the fact that Rembrandt's self-portrait was painted in the middle of the religious fervour of the 30-years war, the decline of symbolism perhaps represented some form of wider secularisation. In this it offers an interesting parallel with the work of the fisherman and ornithologist, Leonhard Baldner, writing only a few years after Rembrandt's picture. Baldner's book is equally poised between two approaches to nature. He extensively quotes the various Old Testament references to the birds which he is listing. He includes with each of the 70 odd entries for individual birds the specific dietary taboos laid down in the *Leviticus* – sample: 'and these ye shall have in abomination among the fowls; they shall not be eaten, they are an abomination: the eagle and the gier eagle and the osprey […] by these ye shall become unclean'. Despite his extensive acquaintance with obscure sections of the Old Testament, however, Baldner in no way allows these Mosaic bans to prevent him from confidently recommending for their flavour many of the birds on his list. Baldner brushes aside the Old Testament prohibitions, explaining that they have been overtaken by the teachings of the New Testament. So he uses a religious argument to carry through a piece of secularisation. Perhaps, Rembrandt felt the same.

The bittern's companion in the wilderness – the pelican – offers a classic example of the intensity with which the birds mentioned in the Bible got interpreted and that intensity offers some explanation for the surprise which in the modern period we feel at Isaiah's unusual pairing of pelican and bittern. It's hard to reconstruct the enormous energies which the early saints and fathers of the church devoted to re-interpreting the pelican; from St Jerome, St Augustine and St Gregory it became usual to see in the pelican not the marker of that doom and damnation which would befall Babylon, but a much more positive figure. The idea starts with the view of the pelican as a hermit, 'living in the desert of this world, sensible to the poverty of this life's pleasures'. Absorbing the legend which believed that the pelican pierced its own breast in order to feed its young on its blood, the interpreters transformed the bird into a symbol of self-sacrificing love, a prefiguration of Christ, therefore, and when the legend went on to claim that the pelican actually killed its own young before reviving them with its blood, the bird became seen canonically as God himself, bringing both death and resurrection to his son and to mankind. And suddenly the desert has become the place of crucifixion, uniting – as the bishop and Saint François de Sales wrote the year before Rembrandt painted his picture – 'solitude and blood sacrifice'. While efforts of this kind picked up its companion in the desert and transformed it into a symbol of God himself, however, the bittern remained untransfigured, hunched in the desert and miserably fishing a livelihood from its brackish waters. As the early historian of ornithology, Pierre Belon, faithfully reported in 1550, the bittern was left with its questionable reputation and became a symbol of extreme social exclusion. The name was used – both by Aristotle and by Belon – as a synonym for laziness. To call someone a bittern, *un butor*, was an easy insult – something which is represented in many of the emblems of the sixteenth century.

Rembrandt's picture shows him in the company of an idler, a symbol of punishment and damnation, a radical outsider – perhaps the same status as artists felt they had. It's reminiscent of Franz Kafka, who responded ironically to his family name, for in Czech *kavka* referred to the crow which was sent out from the Ark to see if the flood had receded but was too lazy to come back and report. The look on Rembrandt's face may reflect the same irony.

There's another source for the bittern's character: the icons of melancholy. The most celebrated depiction was in Dürer's *Melancholia I/II* (1514). A German scholar at the time identified a whole class of creatures which display the features of melancholy, 'creatures of the night, solitary, mournful, brooding, greedy, shy and melancholic' – in short, a portrait of the bittern. The main focus of discussion was the cult book of 1632, Robert Burton's *The Anatomy of Melancholy*, in which many ancient and more modern bird symbols play a part. In Rembrandt's picture the complex of ideas around melancholy cannot be ignored.

The place of the bittern in this complex of ideas is confirmed when, towards the end of the eighteenth century, the renowned naturalist Buffon (see Chapter Five) writes, 'The bittern is as lazy and melancholic as the stork.' In the 1770s the word 'melancholy' is not casually used, it brings with it the range of known elements. So, Rembrandt, no stranger to personal melancholy, is holding a melancholic bird and his expression is perhaps melancholic too.

Culinary considerations

Ask old country folk about the bittern – particularly those living in marshy districts (which is to say before the eighteenth century when so many states began to drain the marshes in extensive parts of the European countryside) – and a very different series of associations is called up. Despite or because of the present day, European legal protection, which the bird enjoys (or at least possesses), country lore reports that the bittern was once considered a delicacy for the table – a 'Fenman's turkey' as an old English phrase suggests. It is recorded that Darwin himself in his Cambridge dining club enjoyed a bittern at one of the club evenings, having other occasions enjoyed owl and hawk. The bittern was by then a self-conscious, snobbish archaism (We recall Darwin sitting at table on the *Beagle* and starting to carve a roasted bird when he realised that his meal was of an unknown bird: the meal went ahead and only the skeleton was rescued for science). However, as we consult various cook books of earlier periods, it becomes clear that the bittern had a much stronger place in gastronomy than it had in biblical exegesis. Baldner was evidently following good precedents in ignoring the various taboos pronounced by Moses against eating certain birds. When it came to the table, it seems anything goes.

The long tradition of the seventeenth-century, Flemish and Dutch 'game-piece' focused on the relationship between the hunt and the kitchen. In one sense, this emphasis was a product of the religious disinterest in hunting – an activity not mentioned in the Bible, though fishing was prominent there – the emperor Friedrich II had struggled to have hunting recognised as 'the quintessence of courtly breeding' but hunting had remained a positive element in Islam rather than Christianity and appeared more often in the kitchen than in the forest. The tradition highlighted the numbers and variety of birds shot rather than particular specimens; they are snap-shots of a whole food chain.

Rembrandt's picture does not stand unambiguously in that tradition. Nevertheless, if we compare this picture with his earlier bittern picture, the *Still-Life with a Dead Bittern* (1637), then it does seem that Rembrandt's work was close to the culinary tradition. In the earlier picture, the bird hangs – looked over by a kitchen maid – next to a hunting rifle and satchel and just above a polished cooking pan and neither picture, regardless of the presence or absence of rifle and pot, evokes a sense that the bird itself is playing any active part. It is offered for its meat, not for its personality. While, as was by no means unusual in the game pieces of the time, the bird is portrayed whole without wounds or disfigurement and the fact that even in the kitchen picture it has not been plucked, does not in itself argue for any interest in aesthetic qualities or detract from its taste.

Digression

Since Rembrandt's picture comes in the midst of a fashion for birds in the kitchen, we'll take a quick look at exactly what role the bittern played there.

A well-known and often quoted example is provided by a massive banquet given for a chancellor of England, Neville, on his appointment as Archbishop of York in 1465. This luxurious state occasion required 204 bitterns and while this figure was modest in comparison to the 4000 mallard and teal prepared for the same banquet, the bittern clearly stood (or hung) at the head of a sizeable industry. Nevertheless, a century later, in a German compendium of menus of 1581 which contains many recipes for banquets of a similar extravagance and details among others four luxurious banquets, 'two of them on meat days', and in addition an Elector's banquet followed by four 'for citizens', the bittern is not mentioned. Could it be hiding somewhere?

We know that it was much too large to be included among the so-called *Kramatsvögel* which were an invariable but apparently minor part of the menus served on meat-days, according to Baldner this group was made up of 'blackbirds, thrushes, mistle thrushes and waxwings'. The bittern was clearly too large to be included in the recipe which required 'every kind of small bird put alive into a tureen'. This luxury dish is the origin of the nursery rhyme about 'four and twenty blackbirds baked in a pie' and singing when the pie was opened. The rhyme suggests another reason why the bittern wasn't included – it wasn't a song bird. There must have been a general assumption behind this, for Blumenbach writes in 1779, entirely without surprise, about the whole phylum *passares*: 'they have tender and tasty flesh and most of them sing'. For once perhaps the bittern's inability to sing kept it out of trouble.

In itself, being large offered no security for the place of honour at the table was occupied by large birds such as the eagle. One recipe instructs the cook: 'pluck it (the eagle) only in the middle, leaving the head, neck and tail feathered, roast it whole and when roasted put it in a dish, so that it looks fine and beautiful'. The recipe radically modifies Moses' ban on eating eagles, which, instead of 'soaring on the wings of an eagle' like the soul entering the presence of God, end up on a dinner plate.

These recipe books are bird lovers' nightmares. The green woodpecker we read 'is good in preserves, pâté or just as you like it'. This is the finer end of gastronomy and when we read in a French recipe that the bittern needed to be cooked with salt only, certainly, as one authority insisted, without sauce, we realise that it is too commonplace for fine cuisine. The 204 cooked for the Archbishop's blow-out meal clearly went to the tables' lower end. In 1384 a bittern cost 18 pence (snipes cost one penny each and for one penny you could get four larches and 12 finches – even a pheasant cost no more than 12) but its size made it reasonable value for money. Was it the burger of the fourteenth century?

By now it is clear why so many ornithological handbooks gave details of what the various species taste like. *De gustibus non disputandum* – it is not clear whether (assuming the question to have any meaning) the bittern did or did not taste good. Rather surprisingly, Baldner not only fails to suggest the bittern for the table, he explicitly does not recommend the bird's flavour, preferring to it among others the Great Northern Diver, though he does offer a use for the bittern's claws, as toothpicks. Laillevant, chef to Charles V, wrote that the bittern 'despite the disgusting taste of its meat to those who are not used to it' belonged to the great delicacies of French cuisine. This judgment was reinforced in 1555 when Belon repeated it and rather surprisingly resurfaced in 1854 in Mrs Beaton's handbook. Something disgusting or a prized delicacy or just for the masses. The ambiguity of the beautiful bird Rembrandt is holding up increases with every new fact.

No sauce but the plot thickens

Such culinary reflections may do something to explain the way in which Rembrandt is holding the bittern in his self-portrait. From the gastronomic perspective, he is presenting it in its plumpness, emphasizing how much meat is on the carcass. Typically, its feet are trussed and it is ready like other game birds to be hung for some time before being prepared for the table. It makes it unlikely that anyone would recognise the bird – upside down is hardly the characteristic *jizz* of the bittern as observed in nature.

Some art-critics have suggested that the purpose of Rembrandt's ambiguous gesture in his self-portrait was not really culinary at all and that he wished instead to demonstrate the wonderful texture of the soft feathers of the underbelly and underwing – indeed, it has been suggested that it is a purely aesthetic pleasure which Rembrandt is seeking to convey – a view which leads to the understanding of this as another of those enigmatic pictures in which the artist uses his ostensible subject to reflect on his own profession of painting. As if the bittern is there merely to illustrate the act of illustrating and by extension the dilemma of art.

Perhaps it is the expression on Rembrandt's own face as he holds up the bird that has led critics towards this interpretation, for the expression on the man's face is hardly that of the hunter. He is not holding up the bittern in the way of a fisherman demonstrating the length of his catch and his face lacks that insufferable smugness which was to mark the countless big game hunters' self-portraits in the later nineteenth century (preserved in photographs rather than by the Rembrandts of the day), in which the slaughtered lions are somehow made to participate in the general satisfaction at a successful shot, sharing a smile with the hunter. It was these which Walter Benjamin had in mind as he commented ironically that the number of illustrated magazines ready to receive such photographic trophies even exceeded the number of game butchers catering for the other form of consumption. In both cases, meat and photography, consumption was the point. Still, we should remember that the element of the hunt is restrained in this picture and if Rembrandt did by any chance kill this bittern himself, then it is hard to think of his pride in his trophy being that of the hunter.

The lack of a clear hunter mentality in Rembrandt's picture need not unduly surprise us since everything we know about the conditions of the hunt in Dutch society of the seventeenth century strongly suggests that Rembrandt – the son of artisans, and although enjoying considerable early success in 1639 hardly a consort of royalty or the aristocracy – would not have been allowed to hunt the bittern himself. Not only did a Dutch treatise on the rules of hunting in 1636 make clear that the bittern was under the 'protection' of being hunted only by the nobility (it is a strange form of protection which determines only by whom one may be killed): the contemporary rules relating to falconry went further and made explicit that the bittern could be hunted only by near royalty. We need think only of the centrality of the theme of the poacher in European literature right up to the last quarter of the nineteenth century therefore to realise that it would have been unwise of Rembrandt to indulge in – let alone positively to advertise his participation in – the pleasures of the hunt. It would have been more or less the equivalent of painting his own portrait on a WANTED poster. Whatever else the picture is showing, it is not the heroicisation of a poacher. He is simply holding a dead bittern.

Between Nature and Still Life

So, we might summarise; here is a portrait executed with a considerable sense of faithfulness to detail in aspects of the bird's appearance, celebrating neither the religious importance of the bird nor the pleasures of the hunt. If any emotion is demonstrated by that sense of detached presentation which is the subject (as well as the style) of the picture, then it most likely to be something closely associated with the kitchen and even then, the bird would hardly have been destined for Rembrandt's own consumption. Perhaps his ironic expression merely expresses the knowledge that he is holding something to which he is not entitled.

In the language of the history of art, this is a perfectly reasonable conclusion. The development of the game piece throughout the seventeenth century showed the increasing luxury of the portraits while the artists' clients were in fact the largely urban middle-class of the newly super rich Holland – a class which had the money but did not have the aristocratic privileges. The glorious feathers of the bittern are, in short, nothing more than a borrowed glory, reflecting more on the purchasers of the pictures than on the birds themselves. In other words, yet another symptom of the excessive consumption which Simon Schama, among other historians, diagnosed.

Reflection on consumption is probably the principal difference between Rembrandt's picture and the kitchen pictures of Frans *Snyders. Snyders, the Flemish painter, regarded as the originator of the whole genre of the game piece, possessed dazzling skill as a still life painter. For both Snyders and Rembrandt, the consumption – like the consumption of the artworks in which they are portrayed – is practised by others – by the rich. When Kenneth Clark remarks that Snyder's innumerable canvases are 'wearisome to us', he misses the point – they are wearisome to Snyders, who remains outside his pictures and refrains from commenting on his non-participation in the feasts being prepared. By contrast, Rembrandt has put himself in the picture and his smile suggests that he is aware of the irony, both of his own position and of his clients, who wish through his picture to borrow the glory of an aristocratic habit. In an enigmatic sense, Rembrandt portrays himself presenting something neither he nor his clients are actually able to enjoy for themselves.

There is – from the point of view of those who think that the bittern is worth portraying for its own sake and that the tastes to which it should properly appeal are scientific and aesthetic rather than culinary – one final interesting feature of these game pieces. That is the way in which the painters within this genre appear increasingly to have turned to the beauty of birds regardless of their taste. One small indication of a possible shift away from the kitchen is the presence of kingfishers, which later painters feature prominently on the heap of game birds. There is a complete absence of kingfisher recipes: Belon, praising the bird as possessing *le plus beau plumage que nous cognoissions*, remarks that not even the peasants eat them – that they simply give them to the children to play with and to enjoy the feathers, so we must suspect that these birds feature in the pictures on account of the colours and the striking and distinctive beauty which they add to the pictures. They were little more than the artist's toy. Perhaps the inclusion of these dazzling birds was a product of surfeit, as Simon Schama would argue, but we must not overstate the case. After all, if the most flamboyant of birds, the peacocks, were pictured quite unproblematically to be eaten (their tail was displayed on the table to lend added glamour to the dish), it seems difficult to think that the aesthetic element of any bird could be important in its own right. Perhaps the artists gradually tired of portraying beautiful birds killed for the table and wanted beauty for its own sake. At the start of the nineteenth century, a Swiss professor of medicine and natural history sadly remarked on this view. The depredation of nature was the result, he concluded, of the 'inexhaustible greed of man, ever plotting the ruin and death of his fellow creatures'.

Something of that lassitude is suggested in the way in which live birds, particularly magpies and parrots are painted in their game pieces, where these live birds stand apart from the trophies and survey the realm of dead nature with the eyes of survivors. We can also observe that artists turned away from human hunting to live animals, initially to dogs but then to wild predators such as weasels and birds. This book is concerned with survivors but it will be many centuries before the survival of birds is felt to be a reason to portray them properly or to restore to life the magnificent birds heaped up as a feast to the eyes. Rembrandt may perhaps represent the emergence of birds into high art, for the birds themselves, however, the road out of the kitchen and back into nature was both arduous and flavoured with their own blood. The legend embraced the pelican but the bittern was left with reality.

Chapter Three
Frisch: Smallest Little Owl (1763)

By focusing the first two chapters on Dürer and Rembrandt I have introduced my subject at the highest artistic level. In the thin air of the mountain tops, only the great eagles of art can circle freely, but I can't claim that either of these artists were over burdened with formal science. It's time to come down to earth with the subject of this chapter, Johann Leonhard *Frisch and his remarkable systematic handbook with the hardly catchy title: *Presentation of the birds of Germany, including some foreign birds, all described according to their properties*. This book was published, following the customary procedure in regular instalments between 1737 and 1763. The title went on to mention its compelling copperplate illustrations in which birds were 'portrayed in their natural colours'. What was really new was that the entire work adopted a scientific style of writing and illustration, something to which the boatman and naturalist Baldner could not aspire. For each bird there is a descriptive section of varying length and a coloured illustration.

It is an impressive, large-scale volume with high quality paper and typography, containing more than 250 full page coloured illustrations. It is striking that, unlike so many contemporary works, it is not dedicated to any ruler or local dignitary. The author was headmaster of a secondary school in Berlin, a proud member of the professional middle classes and he acknowledges his debt only to an institution – the Royal Academy in Berlin and not to the king himself. There's no obsequious preface, simply an 'introductory report' in the name of all the contributors explaining the circumstances and intentions in which the book had come to be written. It reads like the preface to a modern scientific work.

This report gives an account of the 'labours over many years' through which Johann Leonhard Frisch had pursued his scientific work. He had

> put together a collection of most of the birds of Germany. Some he had held caged or had fed on his yard in order to be able the more reliably to study their characteristics and subsequently had them mounted in order to preserve their outer appearance.

This was clearly a more sympathetic procedure than Baldner's, who had justified his inventory of German birds with the words 'I have shot all these birds.' The laboriousness of Frisch's work may perhaps be deduced from the fact that he had died with his work incomplete. Others, notably his younger son, Jost Leopold Frisch, a pastor in Silesia, had to bring the work to completion. The finished work is typical for aiming at the greatest possible completeness in its inventory of birds – in consequence, the preface boasts that the listed birds were 'more numerous' than in the works of its competitors.

The title emphasized the priority which native birds would enjoy. Frisch consciously distanced himself from any exotic temptations, even though he did include some African and American birds. His dilemma becomes clear in his entry on the 'Casuar or Emu', an obviously exotic bird which was later painted as we shall see in the following chapter by *Oudry, on the basis of a celebrated individual specimen. Really the bird had no place in Frisch's lexicon. He admits 'while it breaks our declared principle to write only about birds which were not present in my collection or subsequently mounted, yet these foreign birds have a small claim to a place in these descriptions.' But he goes on in an argument familiar from other zoologists, including Pliny, the cassowary is 'the largest bird in its category and therefore needed to be mentioned'. The argument is strange when one recalls that Europe is the continent with the smallest number of bird species.

In other ways too, Frisch was aware of the limitations of an inventory confined to one specific country, although, for instance in contrast to the English naturalists *Willughby and *Ray, he had not travelled extensively enough to be aware of the European distribution of bird species. Berlin offered a more provincial milieu in the mid-eighteenth century than London or Paris. Frisch envisaged an ornithological science which would embrace the whole world but his real emphasis was on native species, including descriptions of many tamed and domestic birds. For a number of reasons his vision of completeness remains a distant hope. Various species are entirely missing and in particular sea birds are far from well covered. Berlin's distance from the coast may be one reason for that and a glance at a map would suggest the same reason for the shortage of sea birds in Baldner's account.

Names

One of Frisch's principal aims in this lexicon was 'to establish securely the name by which each bird will be known in the German language'. His intention was to contribute to the consolidation of the German language by ensuring that regional words gave way to a unified lexis. Clearly in the early eighteenth century his understanding of what 'German' meant could not be political, for at best the word designated a variety of states for which only the language was (more or less) common. Frisch's principal concern, however, had nothing to do with national identity: it concerned the establishment of a scientific language.

Frisch reserves his most severe criticism for the way in which imprecision in terminology had led to confusion about the identity of species. He is scathing about his predecessors' inability to differentiate between the many species of duck (another problem was small waders). The difficulty was compounded by the terminology used in translations from foreign language books of ornithology. Frisch remarks that his predecessors 'gave ducks their names either according to a bunch of ignorant huntsmen or bits of rubbishy folklore or on the basis of the unsubstantiated opinions they found in ancient books'. The function of the illustrations was the unambiguous identification of the bird in order that its name should be finally fixed and the illustrations had above all to support the text. This was not happening in *Albin's or Gesner's handbooks. Ornithology should be taken out of lay hands and established as a scientific discipline.

As we read Frisch's texts, it becomes clear how flexible the situation of classification was in his time. He was as keen to structure his inventory as he was to make it complete. We shall return to his classification principles in a later chapter but it's clear that the competition was just starting to hot up. Frisch feels himself a pioneer but he has to fight to stay in front. Later, ornithologists would attack their rivals' systems: in his pre*Linnéan age Frisch's problem is the lack of general systems. Gesner and *Aldrovandi put the eagle next to the wren (that is: A next to Z) and water birds were mixed in with land birds. Frisch's wish for an adequate separation into their species and genera sounded more admirable than it was, for Frisch surprisingly still classified bats with the birds.

Despite lapses of this kind Frisch's approach anticipated the position taken up by professional academics later in the century. Blumenbach pours scorn on the 'crowds of dilettantes' who are not prepared 'to come to terms with the abstract teachings on the general characteristics of nature', while Schinz reinforces this view of how his subject can advance as a science:

> Many people find it small-minded if the researcher counts the individual sections of a beetle's foot or describes its colouring or devotes his detailed attention to the tiny differences between closely related animals – yet this is the principal method by which modern natural history has shed light on the history of the human kind, a history which is more and more clearly emerging from the darkness into which the prejudice of earlier centuries had plunged it.

Like Blumenbach, Frisch sets out from a vision of pure science while Schinz reflects on the wider dimensions of the Enlightenment and their cultural and political consequences.

Approaching the Scientific Method

The scientific intention of Frisch's work emerges unmistakably from the individual descriptions. These summarise objectively the bird's physical appearance, its characteristic 'customs' and behaviour, its food and its habitat. Only occasionally, when he is trying to disentangle a bird's name, does Frisch slip into the anecdotal tone which had marked so many predecessors (especially Belon) but that does not reduce his text's objectivity. His methods must have made him familiar with the techniques used to capture specimens but he seldom refers to these and when he does so, it is only because they reveal an important aspect of the bird's behaviour – for instance, whether it moves around in flocks or individually, what food stuffs it finds irresistible and what functions its various calls have. Only in the case of domesticated birds does Frisch mention the flavour of their flesh. This omission was a clear distancing from his predecessors and a refusal to associate his work with the approach of 'ignorant huntsmen' or with peasants' need for meat. Frisch's work was to bear the seal of science: a centralist, expert activity carried out in an urban academy. In his will he bequeathed his collections to the Berlin Academy.

Frisch's scientific credentials are visible in everything he writes, notwithstanding his lack of theory or testable hypothesis. He entirely avoids the speculative thinking still endemic to his subject, on such topics as, for instance, where migrating birds go in winter or about birds' character. And he expresses no opinions whatsoever on the intentions of nature, let alone those of a creator. A hundred years after his work such speculation still padded out ornithological writing but Frisch wants none of it. By cutting out theological reflections or that speculative philosophy the Germans call *Naturphilosophie*, Frisch's work has a surprisingly modern ring without – of course – being in agreement with many of the findings of modern ornithology or moving far beyond the descriptive.

We see this mixture in his opinions concerning hybridisation, the mixing of the species. Frisch's basic position, held in common with ornithologists of many centuries, is that 'bastards' are unable to reproduce or – if they are able – then their offspring are not. Frisch adds a variant to this argument. 'Birds must brood their own eggs – that's an essential part of their reproductive cycle. If another bird hatches their egg, then the chick is a half-bastard.' This sort of claim would, if put forward by other writers, be supported by the authority of an ancient source or with telling anecdotes. But for Frisch, the claim is a scientific rule. The cuckoo – whose young are born 'normal' despite being brooded by a different species – is therefore an exception to a rule. Frisch is aware of virtually no examples of cross species copulation. If the male does not find an appropriate partner, it doesn't become promiscuous, Frisch reports but 'flies from one part of the world to another looking for its kin'. Frisch's theory may be as unconvincing as his observation is deficient (we recall that he kept his specimens, when alive, in a controlled environment, where promiscuity would be all but impossible – he was, after all, a headmaster, yet it is striking that, even when describing the bird's chaste behaviour, Frisch makes no reference to any moral order of nature or to the moral sense of the birds themselves (Buffon took pleasure, as we shall see, in observing and imagining, cross species promiscuity). Frisch aimed at objectivity. He wanted to argue from facts, even when the perceived facts were wrong.

We can observe this feature of his work in his approach to an important issue – one on which Darwin would have benefitted from Frisch's support. Discussions were perennially clouded when it came to the murderous activities of birds of prey. 100 years after Frisch, Darwin had great difficulty in persuading his contemporaries of the brutal facts of the struggle for existence, especially when waged among such beautiful creatures as birds. Darwin confronted opposition both from mid-Victorian sentimentality and from *Paley's natural theology. The objectivity with which Frisch describes nature is striking, as we see in his account of the behaviour of the shrike (*Nomen est omen*: the alternative name was *Würger* – strangler, hardly more friendly than the English 'butcher-bird'). The shrike, Frisch explains, 'does nothing except strangle and rob.' Darwin's illustrators refused to paint this bird together with its so-called 'larder' but Frisch knows no such hesitation.

> The shrike strangles by holding the smaller bird in its beak, crushing the whole throat for so long that it is choked to death. Then it eats the smaller bird but not before it has hacked out its brain and eyes and then takes the head off entirely – it's clearly the shrike's choicest morsel – finally, like other birds of prey, of which it forms the smallest species, it tears all the flesh from the bones […] What it cannot eat at once of its larger prey it hangs on a thorn so that it dies and cannot fall off.

This passage shows not only Frisch's objectivity but a big advantage of his overall method. He's clearly been able to watch the shrike at close quarters – something much less easy to achieve in the bird's natural habitat. Frisch is not the first to make observation his guiding principle but the account of the shrike shows the role of observation within captivity. Frisch couldn't follow the partner-less male bird across the continents but he learned from close observation and it was an observation distinct from personal experience. This method meant that as author and scientist he is hardly present in his text. Observation is a method rather than (as for instance it would be for *Bechstein some 50 years later) a personal accomplishment or life style. Observing nature means that Frisch could make mistakes, but not deal in illusions.

Illustrations as Part of Scientific Progress

Frisch's determination to set up a scientific understanding of nature and to abandon any religion or mysticism – along with his rejection of ornithology as a compendium of countryside wisdom – was nowhere clearer than in his use of illustrations. Unlike so many authors, for whom the collection of suitable illustrations was a second step in their work, Frisch planned them from the start and did so for scientific reasons rather than as a way to make his work more attractive to the market. It gives his book particular importance for our theme.

Frisch describes his concern 'for a true-to-nature picture of every bird' and 'for a careful imitation of its colours'. In this Frisch addressed a double problem which was to plague all successive bird books: a twofold division of labour, first that between the science and the printing and secondly between the copperplate or lithograph and the colouring. At around the time Frisch was planning his work, George *Edwards noted, 'For as long as I live, I am not prepared to give any of my plates out of my hands before they have been coloured. If I do that, they get coloured by unskilled hands.' For Frisch the solution to this dilemma was that another son, Ferdinand Helfreich *Frisch, 'a copperplate engraver in Berlin' was made responsible for both stages of the production.

Despite Edwards' remark it was as unusual for publishers to be amenable to such an arrangement as it was for the scientists to care. On one hand, it was difficult for scientists to keep the publishers in check, especially when it came to illustrations. Many bird books came about as a result of the printer or publisher's personal initiative. A typical example of this practice was the work published in 1749 by Johann Michael *Seligman, a copperplate engraver in Nuremberg. He took his text from Edwards' *National History of Birds* (1736) and the illustrations came from anywhere he could find them, few of his sources being traceable. It was in the spirit of this scarcely regulated market that the translators of Buffon's *Natural History* took over some of the original plates together with others which Seligmann had himself re-engraved and some from Frisch himself. There was no unity of scientific or artistic style. It was that unity which Frisch achieved.

It was unusual, in the second place, for engravers and colourists to work closely together. Colouring was as a rule delegated to menials – assistants, most of whom remained and still remain anonymous. We need hardly add that many of these were women. The practice had hardly improved since the sixteenth century when Gesner's illustrators were at least identified by their initials (in fact, they were wood-block cutters rather than actual colourists). It was, for instance, not until 1990 that the name became known of the colourist who worked for the celebrated ornithological entrepreneur John *Gould in London (and also for William *Swainson). Gabriel Bayfield remained anonymous throughout his working life and for a century afterwards, yet the reputation of those he worked for directly depended on him. A glance at the list of pictures for which he was responsible emphasizes his importance, remembering that every illustration in each copy of a work had to be coloured individually. Gould's wife Elizabeth also worked for years as colourist, gaining a reputation as an artist in her own right. She worked briefly alongside Edward Lear, whose reputation came from his own artistic work rather than through his early activity, which he carried out in brief obscurity.

Downgrading the work of the colourists often meant that the colours in the final editions were anything but true to life. Another major problem was that, by the time the colourists got to see the bird in question (usually as a skin, long after death), the natural colours had faded. This added to the inherent instability of the substances used to create the paints the colourists had to work with. It followed that individual copies of one work could be differently coloured. Frisch was certainly right – at least in the name of science – we cannot comment on his wisdom as a father – when he made his son Ferdinand responsible for both aspects of the illustrations, but it was a huge burden for one person to carry and it is perhaps no surprise that the preface also announced the death of Ferdinand Frisch, after 22 years of unrelenting labour 'tormented by working in war time conditions (i.e., the Seven Years' War) by the inflationary rise in paper prices and in the cost of other materials' (i.e., the substances needed to make up paints: readymade paints did not come on to the German market before 1766). Ferdinand Frisch's 250 pictures represent not merely a milestone in ornithological illustration, they were his gravestone.

His father's ambitions made Ferdinand's work still more of a burden in a further respect. He insisted on jettisoning what had been a standard practice in the profession: the recycling of any more or less useable pictures from existing animal books, including the so-called bestiaries. This process meant, among other things, that Dürer's copper print *Rhinoceros* was reproduced over and over again – i.e., it was re-engraved and published in various contexts, often next to a picture of the unicorn, by people who knew nothing whatsoever about the species itself. It was this practice that made Frisch so critical of discrepancies between illustrations and the accompanying descriptions. For his own work he claims not only the overall quality of the illustrations but also 'that there is no copy here, everything is drawn and coloured from the original' (The phrase 'from the original' similar to the phrase 'from life' could mean from a live or dead bird). That this was becoming a professional standard is clear when *Klein, publishing his own ornithological handbook, felt obliged to reassure his readers that 'the birds mentioned in this work can really be found in nature'.

There's a telling example of Frisch's determination to approach illustrations as a scientific task rather than as a business proposition. It concerns the illustration of a species which we discuss in the next chapter, the so-called King Bird, a Balearic crane. Frisch had himself never seen this bird and felt obliged to give an exact account of the provenance of his knowledge.

> Our illustration once belonged to the celebrated Professor *Ludolff, who obtained it from a most respectable princely library. It then passed into the hands of the late Mr Klein, who sent it to Berlin to be copied. In 1696 a Dutchman had brought a pair of these birds – a male and female – to Ulm, where a skilful painter, Joseph *Arnold by name, painted them and this copy has been made from that original.

Frisch cites the page numbers for the entry in Klein's work and quotes extensively from Claude *Perrault, who had dissected the two birds after their death in the Versailles menagerie. These post mortem reports had been translated in 1757 and Frisch was familiar with them. In his day, therefore, he had had done everything possible to compensate for not having seen the birds himself. This was a clear indication of his determination to be scientific.

Individual Pictures

It's not easy to make a selection here. Some are of the highest quality and even those less successful are striking – and often charming – in their approach to the bird and in the care of their execution. Where possible Frisch produced life-size pictures and, in all cases, gave the exact dimensions of the bird. He also endeavoured to get away from the unsatisfactory nature of other bird pictures:

the static bird perched in the branches looking woodenly to the side remained a cliché of the field well into the nineteenth century. This innovation is best illustrated by Frisch's picture of the bittern. In the background Frisch shows a bird in a characteristic pose, beak pointing upwards, the neck elongated: in the foreground the example is of a bird in the pose chosen by Belon on a horizontal plane while a third bird is shown in characteristic flight. For its time, Frisch's picture anticipates both the work of Josef Wolf and more recent identification handbooks. Despite being dependent on captive or mounted specimens, Frisch places the bittern in a suitable background, a reed-bed but omits any reference to the melancholy which Belon had stressed. The inclusion of the three birds ensure that not even loneliness is associated with the species.

Our principal discussion here is of Frisch's *Smallest Little Owl*. Like Dürer's picture, it's hard to think of it in any other way than as a portrait, though in Frisch's case the picture is similar in style to the majority of the other illustrations. I have chosen this picture not because *The Smallest Little Owl* (**Plate 3**), sympathetic though it may be, is one of his best pictures. Rather it is interesting is to reflect on the effect of Frisch's scientific interests on the quality of the picture and to compare his approach to Dürer's absence of systematic ornithological interest.

Dürer's Little Owl and Frisch's Smallest Little Owl

Comparing these two pictures will hardly seem fair to Frisch. It's no disgrace to come second best to Dürer, but I would like to suggest that the differences between the two pictures are about more than the artists' respective talents or their chosen medium (watercolour for Dürer and coloured copperplate for Frisch). The basis of the comparison is the differing intentions of the two pictures.

Plate 3: Frisch, *The smallest Little Owl*

We can show this initially by considering one of the obvious weaknesses of Frisch's picture. Frisch clearly does not succeed in making his model's eye look natural – we assume it is a mounted specimen with a glass eye (Various critics have expressed doubts about Dürer's rendering of his subject's eyes but there is still a difference). Perhaps Frisch was less sensitive than Dürer but the sensitivity he did possess was clearly dampened by his scientific mission, for the general tendency of his scientific contemporaries was not to emphasize birds' eyes. To do so would reinstate the traditional approach to the 'character' of a species. Character was outmoded and had to give way to a more scientific category. Belon's discussion of the bittern's spiritual disposition did not belong to science. Frisch's notes on the species have virtually nothing to say about character and the bird's aggressiveness is not ascribed to character but to the conditions of its existence. Dürer had no such inhibitions about character, perhaps he even saw the eyes anthropologically as 'the window of the soul'. However that may be, the eyes constitute the first striking difference between these two portraits.

The second difference is between the explicitness of Frisch's portrait and Dürer's less open intention. Frisch's bird is fairly brightly painted, Dürer favours more restrained colours. Frisch thereby emphasizes the markings on wings and breast and on the head – in contrast to Dürer. There is no reason to think that there's a difference of gender or subspecies or age to account for this difference. No such elements are mentioned with regard to Frisch's picture: age, moult and gender are ignored. Some 50 years after Frisch, Johann Friedrich Naumann (who became celebrated for his many portraits of owl species) pointed out how these birds' basic colouring can change radically according to age and season, whereas their markings remain constant. Frisch seems to share this view and for that reason emphasizes the unchanging features of the species. Indeed, that is where he and his father saw the function of their work to identify the constant features of each species. The text above the portrait reads 'Third section of the eighth principal species, third plate. The smallest little owl without ears.' This is followed by a selection of names: 'Noctua minima, s. funerea, le petit Chathuant, s. Chahaun'. The bird is an object of science and is painted as such.

If one claims of a painting that its details outweigh the overall effect, that's a clear criticism of the picture and its painter. The unity of a work of art, the self-sufficiency of the subject count as the criteria of great art: in the context of ornithology, however, things are different. The identification of a bird – it's hard to think of Frisch as a field-guide: the book is too expensive and bulky for that – meant something different in the eighteenth century than today. Without neglecting details, today's bird watchers and their books emphasize the overall impression of a bird, its jizz. That's a reflection of a situation in which birds are not shot or captured in order to be examined and identified at leisure. Illustrators in Frisch's time felt less obliged to show birds' characteristic behaviour and gestures than is the case today and instead concentrated on their unchanging features, present alive or dead.

Indications that this would change come in a study published in 1800 by the important law giver of zoology and botany, Johann Karl Wilhelm *Illiger. In the course of his 'thoughts on the concepts of species and genus', Illiger introduced the word *Habitus*. This did not mean habitat but was defined as 'the embodiment of all the qualities which we observe across all individuals of a species. *Habitus* is really a product of the imagination, an image which the imagination forms of the entirety of a species.' This is an illuminating thought; it helps to explain expert bird watchers' ability to identify birds at distances where it's not possible to tick off a series of objective characteristics. *Habitus* highlights the working together of imagination and knowledge and the idea had clear implications for bird illustrators. Illiger demonstrates that these people must take into consideration the imagination, the way an observer sees the bird. Frisch does not do that.

So, it's not surprising that the non-ornithologist Dürer communicates his subject better than Frisch. Whether or not every detail is right, we can instantly recognise the little owl. Dürer's skill exemplifies the assertion of Alexander von Humboldt when he remarked that the artist's eye is able to penetrate to the inner structures of nature through its power of immediate perception. Classification is a matter of seeing as well as of analysing the differentiation of species, the recognition of the essence of a plant and its affinities with others is for Humboldt an achievement of the eye. It throws into a different light the claim of the great French bird-painter, Jacques *Barraband (and others). 'You cannot draw what you do not understand.' Humboldt's belief was that by drawing one came to understand and Dürer's portrait seems to exemplify that.

We cannot deny that Frisch knew far more about his little owl than Dürer did. We might hesitate, however, despite Humboldt's claim, before ascribing Dürer's achievement simply to instinctive genius and artistic inspiration. Dürer had, as we saw, made a conscious commitment to realism, he set out in his art deliberately to reveal the realities of nature. Frisch too is committed to reality but he is encumbered by the scientific apparatus within which he works and is less good at seeing the whole.

If the newly evolving science of ornithology caused Ferdinand Helfreich a small defeat with his little owl, we might remind ourselves of his achievements in this picture. In the description of the species, the text continues the list of names from the plate by which the bird was known. Among these was a name given it by 'the superstitious mob', the death or corpse fowl (The origin of the name was the belief that if an invalid sees a little owl, then death is certain). Frisch's bright colours are designed to work against this symbolism. That's why the bird is shown in full daylight. The bright colours are no less a sign of Enlightenment ideas than was Dürer's elimination of all traditional religious symbolism. The two portraits are forming their own tradition.

Finally, Frisch's charming portrait causes us to reflect on an issue central to all bird painting, the contrast between what is individual and what is typical in a particular portrait. It's agreed that some celebrated pictures are received specifically as portraits of particularly distinctive individual birds rather than as typicalised pictures suitable for an identification book. This is certainly true of Edward Lear's parrots or some of Josef Wolf's raptors, which he painted from striking specimens in private collections. All bird pictures must be realistic and reproduce accurately the real birds which the artist observes. At the same time the artist has to abstract away from merely individual features and details in their subjects in order to convey the typical, those elements of the species which do not change in their countless repetitions. Science, as Illiger claimed, requires the ability to abstract as well as that of observation. Abstraction can lead to lifelessness or least to a failure to be true to life, turning the bird into a kind of still life, a *nature morte*, but it remains essential to the art.

For all that, Frisch's illustrations – and the book which consumed so much of his family's energies – were more than his tombstone – they mark an early and important phase in a process which aimed to harness the art of illustration to the claims of an exact science. Ornithologists have every reason to respect his achievement.

Chapter Four
Oudry: Painting the Menagerie

Dürer's picture, like Rembrandt's, was intended for the bourgeoisie. Both portraits had a solid and down to earth character, which not even the ambiguities in Rembrandt's picture could obscure. It's too early to speak of artistic realism, still less of scientific method, but it does seem as if both artists had freed themselves (or at least kept their distance) from religious and mythological ideas. Frisch too was proud to represent the bourgeoisie, fully trusting in the truth of science.

Our next picture (**Plate 4**) takes the discussion into a different world, leaving behind Nuremberg, the lower Rhine and Berlin, and brings us to Versailles, the court of the Sun King. It was painted in the years when Frisch's work was about to appear. The three birds are not ones which Baldner could have watched – they add an exotic element to our discussion, but it's an exoticism without fantasy or dreams. No roc or phoenix treads the grass, simply three non-European birds painted with a care for detail which restrains the exoticism.

Plate 4: Oudry, *Toucan, Demoiselle Crane and Crested Crane*

Jean-Baptiste Oudry (1686–1755) was French but his pictures have an immediate connection with Germany. Christian Ludwig II, Duke of Mecklenburg-Schwerin (its capital some 200 kilometres north east of Berlin) commissioned a series of zoological pictures from Oudry. His commission demonstrates the ambition of a German ruler to emulate the Sun King. Even a small state wanted to exhibit the same power as the absolute monarch. Oudry's pictures are one way this transfer could take place and the animals and birds are its instruments. Art historians see in Oudry's work a further weakening of the dominance of history painting, but its function within the monarchical system remained unchanged: to confer prestige.

Clearly a different world but the artist remained dependent on his patron, whether an emperor, king or a rich merchant. Exact science had little contact with the finished product. Yet the social and economic status of nature painting had changed. Oudry's work does not just point back to pre-Enlightenment days, it has clear future potential and some of his animal paintings anticipate Josef Wolf's work in the second half of the nineteenth century. And, as the last section of this chapter suggests, Oudry was close to the work of one of the most celebrated of bird painters, John James Audubon, the French artist who made the USA his home. Oudry is at a cross roads in art history and his exotic birds belong in the history of German bird painting.

Hunting

Like many artists of his time Oudry lived in Paris, hoping for commissions from the king, Louis XV, either for pictures or for the decoration of individual rooms in the palace. Such commissions obviously opened up further opportunities and without the royal favour Oudry's career chances would have been meagre. Responding to these realities, he concentrated on two principal topics. The first was hunting.

Pictures of the hunt belonged in a long secular tradition, possessing no other purpose than to glorify the king. Oudry could do these scenes well, sometimes imitating the kind of medievalism one sees in the Bayeux tapestries. His pictures of falconry are similar, portraying the whole paraphernalia of the hunt: horses, dogs, birds, scenery. His familiarity with this world meant that he got commissions for pictures of individual animals, favourite horses and dogs, such as *Polydore* (1728) – a royal deer-hound – or *Pähr*, Baron Tessin's dachshund. Both of these are portraits concerned to bring out both the physical details and character of his subjects. A contemporary said of Oudry, 'He excels in the depiction of animals' characters, for he has the talent to show their distinctive movement.' This talent meant that Oudry's portraits of animals had the realism which marks his bird pictures and in these also questions of character arise.

Oudry's pictures have much in common with the Dutch still life pictures we discussed in connection with Rembrandt. The Dutch painter *Houdecoeter was a particular model for him, also Jan *Weenix. It's paradoxical to think that the wealthy Dutch bourgeoisie looked to Versailles for an appropriate art, while a habitué of Versailles looked back to bourgeois Holland for his models. Absolutist and bourgeois needs met in the idea of representation.

It's also noticeable that Oudry followed the Dutch masters by painting hunting scenes which cut out human participation, showing instead animals or birds of prey attacking birds, including swans, ducks and a bittern. He had access to the larger predators through the Versailles menagerie, notably lions and leopards and a magnificent rhinoceros. There is also a striking portrait of a cassowary, which emanates violence and power no less than his lions and leopards. The individual bird had a personal history which the staff of the menagerie (and Oudry himself) could not help but know, for on the ship bringing him to Europe this bird had killed the captain, presumably by disembowelling him with its fearsome claws. Oudry's cassowary doesn't only look dangerous, he was dangerous.

In these ways Oudry's hunting pictures moved on from the conventionality of the genre and dealt with violence and aggression in nature. He was, like Frisch, far removed from that nineteenth-century sentimentality which made illustrators shy away from the bloodshed and violence among the 'beautiful denizens of the air'.

The Menagerie

The Sun King had copied many of the legendary despots of the past in building up his menagerie. Anyone reading of China's 'Park of Knowledge' or walking through the Ischtar Gate of Nebuchadnezzar's palace needs no reminder of how the display of fierce wild animals increased the sense of power emanating from the ruler. Another feature of these menageries was to display the worldwide influence wielded by their owner. George III proudly displayed the giraffe presented to him by the Pascha of Egypt, (the giraffe's early demise gave the ornithologist John Gould his career break since he had the job of mounting it) – a display which emphasized Britain's imperial ambitions. Conversely, there are few better ways to mark the fall of great rulers than to read that Nebuchadnezzar was expelled from the company of men and ended up eating grass 'like the oxen'. The ruler of lions became a ruminant.

The link between animal and human hierarchies was most strongly marked in the idea of the animal kingdom. This comes across in many fables and metaphors from classical antiquity to Disney. In the seventeenth and eighteenth century such images fed on La *Fontaine or on *Perrault's fairy tales, which were known all over Europe. Their familiar stereotypes – the sly fox, the vain crow, the improvident grasshopper – all fitted in to the idea of a hierarchy. So deeply engrained were these ideas that the Revolution of 1789 briefly threatened the zoological and botanical institutions of Paris, regarding them as part of the enemy. There were plans to slaughter the menagerie animals for meat and to dig up the *Jardin des Plantes* for vegetables. Buffon's assistant tried to clarify the separation of zoology and monarchism. 'The lion is not a king just because other animals fear him [...] in nature there are no kings.' It is striking how slowly the idea of the animal world as kingdom moved out of the language. *Balzac was one of the few to refer to the 'republic of the animals' but he did so only to satirise the politicians of the Restoration.

For the artists in Paris, just as for the tile makers of Babylon, the link between animal painting and statecraft created a useful market. But the Versailles menagerie had another important dimension, for it encouraged the growth of a scientific academy concerned with zoology (George III too, as an enlightened monarch, wanted his giraffe to have a scientific use). In Versailles it was usual for menagerie animals to be dissected by members of the Academy and the results made available to the scientific community. These dissections had been used by Frisch in preparing the entry for the cassowary in his handbook. For those in the king's favour the menagerie was a chance to see exotic animals and birds at close quarters, alive – if, in the case of the giraffe, only temporarily. Some of the artists we discuss in this book were habitués of menageries, few more so than Josef Wolf and Edward Lear.

The celebrated Swiss illustrator Salomon *Gessner understood that, as well as serving as a permanent archive for dead animals, his pictures could have the same function as a visit to the zoo – a painting therefore as 'a replacement of nature'. The phrase exactly fits Oudry's function for the Duke of Mecklenburg-Schwerin: art replaced nature, for the excellent reason that the Duke could not afford a menagerie of his own. The life size portraits of predators and exotic birds – even a huge picture of a rhinoceros – gave the Duke an enhanced sense of his power and prestige. What makes Oudry's pictures important, however, is how behind the veneer of an almost feudal service to the monarch, they could begin to find space for science.

Paradigm Shift in Natural History

In his influential account of the development of natural history, *The Order of Things* (1966), Michel Foucault postulates a significant change in the discourse of the sciences around 1800. Foucault argues that the centre of gravity shifted away from an obsession with ever more rigid classification systems and towards the 'dynamic telling of the story of the inner development and historical life process (of animals)'. In ornithology, as we examine in Chapter Six, taxonomy remained dominant some way after 1800 but Foucault is referring to the rise of the discipline of ethology as his paradigm shift. Oudry's pictures offer evidence of painters' increasing interest in 'the individual lives of animals'.

Foucault is certainly right. Birds came out of the cabinet of curiosities cherished by eccentric scholars and were gradually integrated into a living natural system. Bird painting documents different phases of this development. It also documents the continued influence of anthropocentrism in the belief that man as the summit of creation should be the measure and standard referred to in birds. We see this notably in the belief that birds' ability to be domesticated represented a positive quality in a species. It was understandable that, in the move out of the curiosity cabinet and in their attempt to show birds as living, painters proceeded in a direction which they labelled 'character' – a concept which could hardly do without anthropocentricity and which constantly risked following the unscientific approach of the emblem books. For that reason Frisch had turned, more or less at the same time as Oudry started his work, against the whole idea of character. But it is also possible to see Oudry's ability to give character to his subjects as a positive step, indeed as not un scientific.

Oudry and Bird's Character

In the Enlightenment important scientific debates took place about animals on topics which have continued – though with different standards of evidence and different focus – into the present. They started with questions about the intelligence of animals, the relationship between instinct and intelligence and animals' ability to have feelings. Ornithologists were acutely interested in these topics, for instance in trying to understand migration and its transmission across the generations. For ornithologists, aesthetic topics too played an important role, not just because of the striking beauty of so many birds but also because of some birds' elaborate and architecturally crafted nests and because questions about birds' intelligence were associated with their 'soul' (and nobody had the slightest idea what 'soul' meant, not even natural philosophers such as Gustav Fechner, who wrote a long and much discussed book on the 'soul of plants), they tended to fall back on the idea of character.

It's not easy for a static picture to make much of a contribution to a discussion of birds' character. A portrait can seldom make clear if a bird is reacting from instinct or experience. A sequence of pictures (later to be called film) is needed for showing how a particular situation comes about, and painters ended up by having nothing to work with except the appearance of the bird as the sole indication of character.

Poor though this was, in many respects Oudry's work building on his reputation for dog portraits brought about a revolution in bird painting, overcoming a rigidity in which birds were little more than coloured silhouettes. The escape from this style did not lie in greater focus on anatomical detail. What Buffon remarked about dogs was shared by our painters. 'The perfection of dogs,' Buffon writes, 'depends on the perfection of their feelings.' He wanted portraits to do more than give an account of dogs' appearance – feeling was more important. Oudry took the first steps on this road.

An important influence was that of Oudry's contemporary, the painter Charles *Le Brun, with whom at one time Oudry shared a studio. Le Brun's place in art history comes from his insistence on the correct portrayal of emotion in pictures. So much of the history painting which had dominated the art world had been (not unlike bird painting) static scenes, full of silhouettes of historical events. Le Brun rejected this sterile art and explored in his own works as well as in theoretical studies the exact expression and gestures which accompany particular emotions. Looking at Le Brun's pictures today we may not be struck with this element, for we are too familiar through news photographs of what strong, not to say violent, emotions actually look like. In Le Brun's day, his ideas were revolutionary. They pointed Oudry in the direction he wanted to go. An early and clear example of this influence can be seen in Oudry's picture of a fox defending itself – the fox's expression shows many features of the studies Le Brun had prepared on the representation of human anger.

It would be an exaggeration to claim this advance as scientific. As far as we can tell, Oudry nurtured no scientific ambitions and Le Brun's reflections on the relationship between the body and emotions are in no way formulated as science. He drew a connection, for instance, between the leopard's 'supple and delicate body' and its 'sly and deceitful character', while the bear with its 'coarse, wild and fearful body' had a 'cruel' character. That's not much more scientific than the La Fontaine fable of the bear who tries to help his human friend in the garden but ends up killing him out of clumsiness. La Fontaine's is funnier, but neither of them can lay any claim to science.

All this will sound banal, hardly worth a footnote in any history of zoology. In 1872, however, when Darwin published *The Expression of the Emotions in Man and Animals*, these topics moved centre state in scientific debate. In this book Darwin concentrated on the similarities between the expressions which articulate particular feelings in animals and humans. Using pictures and photographs – among others from Josef Wolf – Darwin examined what had fascinated Oudry in his picture of the fox, just as it had preoccupied Le Brun before him. The topic itself implied a final answer to the old debates about whether or not animals had feelings, for it could be assumed that if animals had access to the same expressions as humans, they must experience feelings and therefore have character.

It can be misleading to use a more modern scientific understanding to emphasize the modernity of older theories. Modern atomic physics did not prove Plato's theory of atoms, but it would be wrong not to see Darwin's enquiries within the same context as Oudry's pictures. The mediation between the two was certainly far from direct or respectable, having passed via *Lavater and physiognomy regarded today (and by Darwin) as a pseudoscience. Oudry is at the beginning of a faint but real trail to science.

Three Birds Full of Character

We can finally come to Oudry's *Toucan, Demoiselle Crane and Crested Crane*, painted in 1745 for the Duke of Mecklenburg-Schwerin. It's a large picture, the birds are nearly life size. We should not see the picture immediately as an example of Oudry's 'careful naturalism'. However, much Oudry tried to place the birds in nature, a clearing in European woods, his picture remains highly artificial and does not escape the menagerie.

The two cranes were well-known figures in Versailles. They strolled around the place much as the royals themselves, indeed they had become one of the trade-marks of the place, like the panda bear for the WWF.

We need to consider the species Oudry has chosen to paint. Admittedly the cranes come from the same continent (albeit from widely distant parts of it), but the toucan came from South America. The toucan's first European showing was in Gesner's *Historia Animalium*, but his appearance in this picture was a chance event, owing most to the colour combination which allies him to the cranes. It would – even without thinking of the species' future role in Guinness advertisements – be futile to look for evidence of his character: he's just an extra.

The two cranes, however, are stylised, emphasizing that link between anatomy and character which was Le Brun's starting point. They stand like monarchs with their crowns. Their demeanour indicates aristocratic arrogance. In the emblem books cranes had symbolised watchfulness, standing on one leg while the other held a stone which, if they fell asleep, would fall on the other foot and wake them. For his part, Aristotle regarded the cranes as dancers, while Perrault and the other dissectors compared the 'Numidian maiden' with 'dancing gypsy women'. Oudry remains true to his commission and paints the cranes not as dancing gypsies – who would not be admitted to court – but as members of the royal *corps de ballet*. The representative function of the picture outweighs antique symbolism.

Oudry clearly had many opportunities to observe these birds from close quarters. That's indicated by the detail with which the crests are painted. Only studying 'from nature' makes such detailed work possible, but we can see in this picture influences from other painters, in particular from the tradition of formal architectural paintings of which the English illustrator Francis *Barlow was a pioneer, but which more recently had been carried on in Versailles by the Flemish painter Pieter *Boel. Barlow places his birds ornamentally before the façade of classical buildings. Oudry's picture, for all its zoological care and the forest setting, shares something of this formal aesthetic structure.

For that reason, Oudry, despite his progress in depicting animals' character – remains a painter in the Rococo style. The realism of his picture is deceptive but his work draws attention to a further aspect of the activity of many nature painters, including those who were to work on illustrations for Buffon. Oudry's commissions came, more regularly than those from king or aristocracy, from the Beauvais tapestry factory and from the famous porcelain factory in Sèvres. Part of his work went directly into industrial production. This mixture of aristocratic and industrial patronage was typical for many nature artists, not just in the eighteenth century. Martin Kemp gives the example of the educational reforms after 1806 in Prussia, in which the state built into the curriculum elements of industrial design, a kind of bridge between art and industry. As late as 1874, *Flaubert illustrates the situation in his novel *L'Éducation Sentimentale,* where he shows the tensions between idealistic artists trying to make their way and

Jacques Arnoux, an important entrepreneur in the art world through being the proprietor of the periodical *L'Art Industriel*. What, for all we know, was a harmonious relationship for Oudry was for many artists a real dilemma torn between pure and commercially exploitable art.

Oudry's career illustrated an acute problem for flower and bird artists – an almost irresistible pull towards the decorative. Since Dürer's time, these art forms had lacked a utilitarian element, so that serious artists were tempted towards the luxury market. Extreme examples of this are seen in Audubon's prohibitively expensive folios or in Redouté's roses. These were the pictures which would inspire Gobelins and expensive porcelain. Another gifted painter Pauline *Knip – who trained with Jacques Barraband – was forced to divide her time between scientific work for Coenraad *Temminck, director of the Leiden Museum, and commissions from Sèvres. Historians are right to put Oudry in the Rococo school, but within bird painting his decorative style represents a characteristic challenge to a more scientific method. Notwithstanding this tension, Oudry managed in a crowded field to produce pictures which deserve a place in the history of the branch and to have anticipated that paradigm shift in natural history of which Foucault wrote.

Oudry and Audubon

Perhaps the first name people know in bird painting is that of Audubon. To include him in our survey has a justification similar to that of Frisch including foreign birds in his list of German avifauna: that he was too big and too obviously among the leaders of his category to be left out. Audubon is often seen as a pioneer and innovator in bird painting and many critics have thought of him standing outside any tradition. 'Born into a continent without museums and academies,' we read, 'the American artist embraced nature as his only teacher'. This is emphasized in Audubon's biography by his unambiguous rejection of eighteenth-century ornithology. It was reading Buffon's error strewn account of North American avifauna which determined him to write his own handbook and to get away from the fantasies and untruths he found in Buffon. Like the early European ornithologists, Audubon set out to draw up a complete inventory of 'national' birds – his project was scientific in nature and culminated in *The Birds of America* with more than 430 gigantic pictures, in which Audubon depicted more than 1000 different species. The dynamism of these pictures is not diminished by the fact that he invariably shot his subject first before arranging them on wires in the postures in which he painted them.

Audubon's reputation was and remains founded on this work, if rightly challenged elsewhere. The few critical comments from scientific ornithology to the effect that he had left out telling details or been guilty of incorrect identifications have done little to shake that reputation, but it is possible to approach him in a slightly different context, not simply as a backwoodsman and pioneer of *plein air* bird painting but as a glorious continuation of European tradition and here Oudry has a modest role to play, not for the dynamism of his pictures – though Oudry was capable of dynamism, less in his bird pictures than elsewhere – but in the decorative element common to both.

With this in mind, we turn to our final picture, the *Dead Crane* (1755 – **Plate 5**). It's not just that Oudry shows the bird in an inverted posture much favoured by Audubon but he completely integrates the body of the bird, its proportions and colours into the natural background. Setting and bird together become a striking decorative ensemble, and this is the only way to describe so many of Audubon's pictures, for instance *Storm Petrels*, where the twisting bodies of the birds and the configuration of their feathers are perfectly integrated with the lines of the dashing waves. Audubon's picture is more dynamic but is constructed on the same principles as Oudry's. Style is set exclusively by the bird.

This dynamism makes it hard to speak of Audubon's work as still life but his pictures are decorative and even ornamental. There is no model for this in American art, certainly not, for instance, in the animal pictures of John Singleton *Copley. Oudry as a characteristic nature painter of the pre-Enlightenment is closer to Audubon than one might at first imagine. After all, Audubon did his earliest bird studies during a long stay in France at the turn of the century. His apprenticeship to the painter Jacques-Louis *David offered one route for this mediation. Audubon's spectacular bird paintings did not start from nowhere, nor simply from confrontation with the raw, untamed nature of North America. The idolisation which he received from Europe – the Germans saw him as 'a man of the forests and savannahs', while Chateaubriand enthused 'there is nothing old about America except the forests' – was a typical Romantic misreading. If his work embodies the size and energy of the new continent, his art stayed within some of the confines of the old world. The huge project which he represented in bird painting extended European achievements without leaving them behind. Yet his pictures, taking over the ability to represent an aura which is more than scientific, do not celebrate the power of a corrupt monarchy – they give symbolic expression to the limitless power and energy of the new continent, its nature and its birds.

Plate 5: Oudry, *Dead Crane*

Chapter Five
The Last Great Natural Historian: Buffon's Natural History and Its Impact on German Ornithology

The last chapter showed behind the ornithological detail of Oudry's pictures both a decorative tendency and a renewed interest in birds' character. The decorative tendency was allied to the ambition to demonstrate in animals the majesty of human rulers. Oudry's concentration on character, while it was coloured by the monarchy, possessed a weak link to the growth of ethology.

We turn now to George Louis Leclerc, count of Buffon (1708–88) and his massive work of natural history. His theme is the whole of the natural world, the section devoted to birds comprises nine volumes and 42 instalments, appearing from 1765. He belongs to the age of the so-called encyclopaedists. His work's size is daunting, its importance to natural history unrivalled before the mid-nineteenth century. It cannot be fully dealt with here. It confirms my central argument (that bird painting raises major issues as it passes through the centuries) but I wish neither to trivialise Buffon by claiming to have discussed him thoroughly, nor to trivialise my theme by leaving him out. I am aiming at no more than to present Buffon as he affected bird watchers and illustrators.

After the revolution of 1789, the Terror left deep scars on subsequent generations of scientists and pre-revolutionary figures did not have a good press. Buffon's son, whom he had apprenticed to his colleague *Lamarck, went to the guillotine in 1794, while in Germany, though the Terror did not cross the Rhine, the shadow of violent revolution hung over many scientists and intellectuals. It was used as a justification for repressive policies by the German states and as a warning against scientific radicalism as when the celebrated scientist Rudolf *Virchow warned that *Häckel's Darwinism would usher in bloody revolution on the streets of Berlin. (As a result of such fears Prussia removed biology from the school curriculum in 1882.) Throughout the century, however, Buffon remained the target of radical scientists – in particular Carl *Vogt repeatedly attacked the slave mentality which he believed Buffon to have encouraged among scientists.

Despite these reservations, if we think of the double function of the royal menagerie – to enshrine power and to further scientific enquiry – it offers a convenient way to differentiate between Oudry and Buffon. Oudry worked at representation and remained only superficially touched by science. Buffon worked at science and was only superficially touched by representation.

Buffon and the Science of His Time.

This differentiation is defensible but doesn't imply that Buffon's work reflected all the progressive tendencies of contemporary science. His life's work is a testimony to a man who loved nature and felt compelled to collect and catalogue it but his at times child-like enthusiasm left out important dimensions of what science is.

The most modern feature of his work was its insistence on observation. The emphasis on empirical knowledge marked the whole Enlightenment and Buffon was its child. His writings display both his own gift for observation and the importance he attaches to other people's observation. He repeatedly cites other sources, not classical authorities but his own contemporaries, who report from the field. Clearly, he did not devote sufficient attention to anatomy, or, for instance, to the structure of a bird's plumage. Nevertheless, without his example, German ornithologists would never have found the courage to build up their subject.

Buffon's influence on German ornithology is epitomised in a handbook published in 1791 by the youthful Johann Matthäus Bechstein, once called the father of German ornithology. In the preface Bechstein admits that he 'has followed only my own experiences' and that neither the hope of academic success nor any theory had led him to put pen to paper. The justification for his book lies in his experiences as a field ornithologist – 'Indeed, I dare to claim that there is no bird in Germany or at least in Thuringia, which I could not recognise at some distance, both from its song and its flight.' That sounds a modern ambition, but in fact it came from Buffon, who may not have been the most scientific ornithologist but who gave younger ornithologists a confidence which few other scientific works could match.

Not just Buffon's lack of interest in scientific anatomy but his refusal to engage in systematic classification would not find favour with modern ornithologists. He accepted the traditional divisions of avifauna but paid scant attention to details, for he was interested in the individual specimen rather than the species. Buffon rejected the work of Linné but made no efforts to promote a different classification system. He held fast to a traditional belief in the indivisible unity of nature, the idea of the 'great chain of being' (the *scala natura*). For Buffon this idea indicated an unchangeable, not artificial structure in nature.

Buffon's rejection of Linné had nothing to do with differences of opinion on matters of religion. Linné was probably more God-fearing than Buffon and certainly believed no less firmly in the divine unity of creation. In 1758, however, following some strange logic, Linné's work had been put on the Index. Some German scientists attacked his system on the ground of morality, rejecting Linné's classification of plants according to their reproductive systems – a leading botanist, *Siegesbeck, regarded that as an insult to the Creator. While this opinion briefly put a brake on sensitive souls' enthusiasm for painting flowers, Buffon was unimpressed by such moralism. Indeed, he identified an even greater range of roles for sexuality in nature.

One might see Buffon's rejection of Linné as a trivial difference of opinion, not even touched by religion – in short as no more than pedantic jealousy. Yet the issues anticipated important debates in the nineteenth century. As the promotion of the exact sciences led to the breakup of established areas of knowledge – starting with the breakup of 'natural history' into a whole series of separate disciplines – so it became easy to talk of the fragmentation of knowledge. This fear allowed ideologies which laid claim to a holistic universalism to present themselves as rallying points for opposition to 'materialism'. The widespread popularity of 'natural philosophy' at the end of the eighteenth century drew on these fears and, as we shall see, the zoologist Lorenz *Oken played an important role in mediating the ideas of the Romantic philosopher *Schelling to an age fearful of science. Increasingly, not just in Germany but in England and America too, these ideas became a source of hostility to reason itself. Buffon himself was non-ideological in his opposition to

Linné. His work has most to say of his love of nature and his empiricism is stronger than any ideology.

In this chapter we are, once again, discussing German developments by reference to foreign influences. It would be possible to see this in two ways. The first would be to note that Germany lagged behind France and England in terms of the general institutions of modern urban life – it lacked the centralisation from which Britain and France profited. So, Germany had hundreds of courts, a plethora of provincial capitals, several academies, dozens of universities but few as powerful and as outward looking as the *Académie* or the Royal Society. In terms of natural history, it had few enough figures who had travelled the world (though Georg *Forster became famous for his account of having travelled with Cook) or addressed a European audience. Alexander von Humboldt had been a lone exception and it was no coincidence that on returning from South America he took up residence in Paris. A more positive view would be to stress the openness of German scientific life to foreign influence. Either way, Germany needed big figures and Buffon met a need.

Two Translations

We shall focus on two major German translations, both of considerable scientific importance. The first translation started to appear in Leipzig – the centre of the German book industry – in 1772, over the name of the translator Friedrich Heinrich *Martini, the second shortly after that in Berlin – here the translation was in the hands of Carl Joseph *Oehme. Buffon's work was not completed when these two publications started to appear, his work was already being anthologised in France and the German publishers were trying to cash in on a significant scientific event as it happened and to milk the success of a major best seller.

The translations communicate this sense of urgency. Buffon represented a dynamic and marketable side of science. It was important to keep up to date. If an individual instalment was late arriving (a common occurrence with this form of publishing), then elaborate excuses had to be offered. In France, Buffon was in the same situation, apologising if an expected volume was delayed. This mixture of commercial and scientific marketing marked the nineteenth century, common to both was the orientation towards the reader. As Loveland points out 'Eighteenth-century readers read natural history not just for knowledge but also for pleasure, distractions, entertainment.' If may seem frivolous to focus discussion of Buffon on illustrations but it may be close to the spirit of his intended readership.

The notes and additional material mentioned on the title pages supplied up to date references to the scientific literature. They also served to harmonise Buffon's bird list with the range of birds to be seen in Germany. These were scientific tasks and Oehme wrote a long preliminary text under the title *On the most important ornithological books and systems*. As he placed Buffon's work in current scientific ornithology and its methodology, he was critical of Buffon's methodology and especially his lack of interest in systematic classification. He can agree with Buffon in dismissing systems which classified birds according to a single anatomical detail – his words are directed against Klein and *Brisson, who made the arrangement of the toes the sole basis of their systems, but he praises Linné explicitly and at length. Linné had introduced a natural rather than artificial classification, Oehme writes, 'Linné manages to integrate so many features into his order that one believes he is copying nature itself rather than setting up an arbitrary system.'

It was clear that the royal menagerie gave Buffon's work much of its order. His inventory starts with the most powerful birds and after running through the principal species, Buffon explains that the birds which come next will be those 'who have an affinity with the eagles'. In fact, they are 'foreign' birds which land in that section. Apart from these divisions, Buffon is not interested in finer differentiations between groups of species. An 'affinity' is enough for him. He's careful to start each volume with an eye-catching bird, not with any new taxonomic order. Yet, despite the royalist origin of his thinking, the non-royal birds, the commoners, so to speak are treated with full seriousness.

Buffon attaches great importance to bird names. Unlike Frisch, he argues for the recording of as many names as possible, in various European and non-European languages. He wants these names to receive proper attention since, were names to get lost, cross-border confusion would be inevitable and valuable information about the birds' behaviour lost, particularly in the case of migrating birds.

His German translators had a different approach to names. Oehme insists on Linnéan classification, although he keeps the lengthy footnotes in which Buffon follows up on names, while Martini spends more time preserving names, indeed adds both the Latin and Greek names of the species. Martini regards the ornithologist as a member of the community of scholars, a man of books. Bechstein understood Buffon better when he identified ornithology as a subject of field observation. We shall see this difference played out in the illustrations.

The English Translation

The complete English translation was later to arrive than the Germans. It lacks their sense of urgency. The delay was caused, according to the 'Preface by the Translator', by the 'great expense that had long prevented Buffon's Natural History from appearing in English dress'. Mr Smellie of Edinburgh had begun with the translation of the first part but he had 'not chosen to complete the task'. This was a matter of regret since Buffon's work means both scientific advance (i.e., new knowledge and new methods) and the advance of science as an institution. Mr Smellie's translation, we are informed, was 'addressed to a narrow circle of readers', whereas 'the chief object of translation' is to accommodate 'the numerous class of readers'.

In pursuit of the first aim, the translator follows his German colleagues in offering extensive additional notes on points of science. There are no footnotes in these texts but the help of the ornithologists *Latham and *Pennant is acknowledged. The same people assisted him in correcting 'a few inaccuracies'.

The English translator shares Frisch's dislike of any confusion in nomenclature but is less bothered with arguing about classification systems than Oehme. Times have changed, in the sense both that Linné is more established than before, and that debates on classification have become more frenzied. The translator's view is pragmatic. Any 'arrangement, however, artificial soever, helps to bring order into natural history. He himself is clearly a Linnéan. He writes, 'To complete Natural History requires the union of Buffon and Linné.' He rounds off his fulsome praise for Buffon with the claim that, 'His lofty genius burst from the shackles of method' – Oehme was less clear that there was any method.

While neither Martini nor Oehme had any sense that translation involved cross-cultural awareness, Buffon's English translator is preoccupied with such matters. He claims that Buffon's 'eloquence' was what 'first rescued Natural History from barbarism' – he means by this the clarification of the discipline, having been a science 'cramped by artificial systems, encumbered by coarse and obscure jargon and disfigured by credulity and ignorance' – yet he has reservations. He retains an elevated style, yet he has had to 'abridge and condense' the text, removing those 'sprightly turns in French which the masculine character of our language will not admit'. He clearly knows that similar sensitivities inform his readers and – since the best thing a translator can achieve is 'to avoid censure', for there is 'neither honour nor emolument in the labour' – he chooses to remain anonymous.

These are but diverting glances at the text. In comparison to the approach which Martini and Oehme take to philosophical issues in their subject, the English translator approaches them from the point of view of Paley. He justifies the subject in terms of its ability to reveal 'the most enchanting prospects of that wisdom and power which upholds and conducts the universe'. Buffon's sceptical attitude to the intrusion of religion or morality into his subject has not impressed his translator, who finds such attitudes part of 'a certain superficial craft, not altogether suited to the manly sense of the British nation.' Again, we see the difference between English institutional thinking and the highly secular nature of French and German academic life.

Illustrations – More than 'Bad Painting'

The English translator mentions the illustrations primarily as a cost factor, but for Buffon and his German translators they possessed a high commercial, scientific and artistic importance. Buffon describes the pictures as 'the best in this kind of bad painting which we call coloured illustration'. He emphasizes the colours, explaining that in the field 'the colours are crucial, often the only features which distinguish one bird from another'. He refers to one courtier who – on the basis of black-white illustrations – thought that all animals at the North Pole were white.

His comment points to the refusal of the art world to regard as art any pictures with a scientific intention more explicit than Oudry's. Buffon is conscious of the great progress made in illustration, mentioning the work of Frisch, *Martinet (Buffon's engraver) and *Daubenton himself (editor of the *Planches Illuminées*) and stresses that all illustrations are based on copying nature, although it wasn't quite clear what the phrase meant. Only a few illustrations are labelled 'From Nature' and these are mostly those which show an anatomical detail, such as a bird's respiratory or digestive system. The dissection room had become a kind of second nature. Few illustrations show the birds in their natural habitat, which was probably why Buffon called illustration 'bad painting'. Other painters always put in a background, often an inauthentic background, and that meant that their work counted as 'art'. Their pursuit of science caused them to be downgraded as artists.

Buffon's translators outdo Buffon in their concern for the illustrations. It was not an easy task, particularly since neither Martini nor Oehme could command, as Buffon did, a staff of 80 to handle the illustrations alone. In the introductory discussion of the 'most excellent ornithological books', Oehme's principal criterion is the illustrations. He directly addresses the issue which Buffon referred to when he spoke of illustration as 'bad painting'. Oehme notes that many illustrators overemphasize the aesthetic element and complains that not just in the illustration of exotic birds, 'the birds are painted too beautifully'. It's a mistake often made with well-known local birds.

Buffon's illustrations and those carefully selected by his German translators are characterised by scientific stringency, a complete separation from the fanciful pictures of earlier traditions. They did not just take over pictures from Frisch, they adopted his principles, incidentally opening his work to a far larger audience than it might otherwise have had. Buffon's translators gave no commissions, they merely selected from what was available, even surprisingly archaic works.

The editors' problems were very real. In addition to the problems Frisch faced (which had subsided with the end of the Seven Years' War), there were new ones. While Oudry simply produced his original paintings and sketches, Buffon and his translators had to solve the problems of reproduction. New plates had to be engraved when pictures were taken over from other sources. (Sometimes the name of the second engraver is included with the picture.) Then the colouring had to be organised. The further in time the original engraver was from the second engraver, the greater the author's physical distance from the colourists, the greater the danger of errors. Cheap editions needed no colouring but for the expensive editions the illustrations needed to be coloured individually. Each volume had about 70 illustrations, and the co-ordination was the responsibility of the translators and caused more problems than their scientific editing.

Martini and Oehme took over three quarters of their illustrations directly from Buffon. The remainder came from contemporary French and German sources – often the same sources which Buffon had used. Frisch and Seligmann were the most important German sources, and it is interesting to note the place given in the scientific literature quoted to non-European sources.

The Frontispiece as a Key to Buffon's Reception

As one would expect in a prestigious project, both translations were provided with an appropriate frontispiece. As the publications continued through the years, the printers (or editors) chose different frontispieces – as one might expect, often taken from familiar models – and the sequence of these pictures reveals the thinking and purpose behind both projects.

We recognise their basic ambition from another contemporary source: the preface to the collection of translations from the post mortem examinations carried out on animals from the Versailles menagerie. The anonymous translator explained his intentions as follows:

> In view of the fact that in our days science does not remain confined to the lecture-theatres and studies of academics, but increasingly finds its way into the studios of artists thirsty for knowledge, into the clerks' room of successful business men, and into the homes of other persons endowed with natural reason – indeed, science finds itself on the bedside tables of enquiring women, in short it finds its way to all sorts of people whom one might not expect to have any such inclination to scholarship – in view of this, it seems not inappropriate to make this particular work more widely available to these people in their mother tongue. A work which – because of its expense – has been hardly available even to many scientific researchers.

The ponderous language may not seem especially appealing but the determination to exploit popular interest in science could not be clearer. For a more substantial work, such as the volumes of Buffon's bird studies, the orientation to the reader – what Loveland refers to as Buffon's 'vulgarisation' of natural history – was of central importance and it's rewarding to follow how the frontispieces reflect that orientation.

Martini selected for his first volume – that is to say, the first volume of the first translation – an expensive coloured frontispiece by a celebrated artist, Daniel Nikolaus *Chodowiecki (**Plate 6**). Chodowiecki was at the time at the height of his fame. His aesthetic creed was naturalistic, his motto 'nature alone is my teacher' but his view of ornithology archaic, following iconographic tradition rather than reality. He depicts a collector in his study examining a mounted bird, which we assume he has just acquired. What is striking is how old-fashioned the collector and his study appear, how out of keeping with the direction which science would take. There's no indication of laboratory work, for instance, and still less of any field work (indeed, the collector seems a little old to venture safely outside). In his night-cap and dressing gown, the ornithologist looks like the hero of an E.T.A. *Hoffmann short story rather than a professor of zoology.

Plate 6: Chodowiecki, *Untitled frontispiece*

The public to which this picture would have appealed was small and regarded science as a leisure activity. It's even less up-to-date than the idea of a scientific library, such as the library of the Duke of Weimar on which both Goethe and Oken depended for their work. (Oken greatly irritated Goethe by taking books home.) Science is portrayed here as the property of a narrow, educated elite. So unsurprisingly this frontispiece was changed.

The next frontispiece (**Plate 7**) dates from 1775 and also fails to express a new spirit of science. The scene is reminiscent of the old hedge schools. Protected from the wind by a sheet hung from the trees, two boys draped as in classical art are drawing birds. One holds up a drawing which, we presume, the other boy is endeavouring to copy. There's a pile of unsuccessful sketches – the boy must improve. So, we have a classic Enlightenment scene, free learning in and from nature, the arts practised in harmony with nature. The publishers clearly felt that this world-view would secure readers since it epitomised the ideals of an age which had taken Rousseau's *Émile* to its heart. It did not seem important to stress the book's more narrowly scientific credentials.

Plate 7: Unknown artist, *Untitled frontispiece*

Before the next frontispiece (**Plate 8**), there had been serious rethinking, for the basic concept changed. We see an adaptation of a conventional hunting scene. A man is hiding in the bushes by a lake. From his hide he can see a large number of ducks and high above a flock of geese. It is framed as a typical scene of country life, such as Baldner came from and which Johann Andreas *Naumann would reflect in his handbook (cf. Chapter Seven). It seems as if the publisher wishes to attract countrymen to become purchasers. A closer look, however, reveals that this is not a conventional hunting scene. The tube which the man is holding is not a rifle or shot-gun, but a telescope. He's holding it not to his shoulder, but to his eye. If he were to fire a gun in that posture, he would knock himself out. If we compare the length of contemporary hunting guns, we can feel certain that the picture is showing something else, in fact killing is being replaced by observation.

Plate 8: Unknown artist, *Untitled frontispiece*

There is more evidence of the subversion of conventional hunting in the next frontispiece (**Plate 9**). A horseman is passing a falconer, whose attention is focused on his falcon, which remains hooded. The horseman pays him no attention, for he is watching a drama played out in the skies as a wild falcon has just struck a passing heron. The horseman's interest is captured by nature's spectacle – he does not even glance at the huntsman. It's a hunting scene in which the excitement lies with observing the natural spectacle.

Behind this progression is a search for positive images of biological science, something which that science continued to lack, even while the educational policies of Germany shifted towards the technical industrial state of the later nineteenth century. Other new sciences quickly found images with which they could be positively associated – the biologist and chemist bending over their microscope and test tube, the engineer standing proudly in front of a gleaming locomotive or *Menzel's *Iron Rolling Mill*. But ornithologists lacked such images. Perhaps the time-lag can be explained as natural history's resistance to its modernisation. Buffon's illustrations represent a turning point for publishers and illustrators alike. They would have to stop building on a traditional public – their future strategy would have to harness their interests differently and to find a new public with the assistance of science alone.

Plate 9: Unknown artist, *Untitled frontispiece*

The Future of Ornithology

Buffon has a vision of his subject's future. He wants it to become a science in itself. That he was opposed to systemisation is no more than an apparent contradiction for much of the classification of his time was at best rudimentary and formulaic. Linné's work was on a different level altogether, which Buffon should have recognised, and Buffon was over-dogmatic when he asserted that 'Nature does not tolerate separate divisions'.

Buffon is aware of various factors which inhibit the development of his science. He felt overwhelmed by the numbers of unknown birds sent in from ever more remote parts of the world. 'Is there any hope that we can collect them all?' he asks in near despair. He was a typical encyclopaedist but should have taken comfort from the way botany had coped with a similar problem. While classical authors knew of some 500 plant species, by the eighteenth century that number had risen to over 20,000. Yet botany had survived and Buffon hopes that his subject will do the same while being aware of its particular difficulties, notably the need for each species collected to include male and female adults, juvenile and winter and summer plumage, that is at least five birds.

Another problem was that his subject was concentrated on Europe and although Buffon had correspondents in Madagascar, Senegal and Réunion, there were countless birds from other parts of the world which he could not identify. This was why ornithology was not able to solve its greatest mystery: migration. Without observers in all countries, one could not even be certain that migrating birds actually went somewhere else. That other animal species hibernate in winter had offered a tempting analogy, and the theory that migrating birds buried themselves in marshes until they re-emerged the following spring had many supporters. Buffon is not persuaded by this explanation – he wants ornithology to be based on observation and evidence, not on speculation, but he is painfully aware that the subject is still too little developed for this to happen.

On the Nature of Birds

It would be unfortunate if we concentrated so much on translations of Buffon that we missed the pleasure of hearing his own voice. It was that voice which made his work so beloved beyond scientific circles, even if it could mean that he was taken less seriously as a scientist. Not only had his revolutionary critics accused him of creating *belles lettres* rather than science. The late eighteenth century was characterised by the search for a popular and understandable scientific style. Buffon tended to look backwards to classical models, although his wit and directness meant that the style never became mannered. In future, scientific rhetoric had to distance itself from aesthetic quality. Such a view comes from Blasius Merrem (professor in Duisburg) who admitted of his own writing that readers would miss out on 'the agreeableness which I should perhaps have lent my style'. But he feels justified for 'I thought I was writing for scientists, not amateurs'. Buffon's style was clearly in danger of being regarded as something for the 'amateur'.

His style is shown nowhere to better effect than in the introductory chapter which informs his readers about the nature of birds and his own love of them.

For Buffon, birds are physically beautiful and their behaviour engaging. He is entranced by the love of freedom expressed in their flight. He admires their family life, the sociability of their homes. In all these dimensions Buffon sees great affinity between birds and humans, indeed 'the tenderness is half envious'.

Buffon's gets close to sentimentality at various points. He, an admirer of the absolute monarchy of France, regards bird of prey as 'tyrants of the air'. He is aware of the usefulness of birds, their practical value to humanity. Like Darwin he uses domestic fowl to draw conclusions about wild birds, yet their value is not purely utilitarian but moral too. The chapter is full of examples of the various species' virtues. Birds are 'a serious and decent race, which exhibit excellent lessons and laudable examples of morality'. (In the original, Buffon is referring to the morality which features in *Aesop and La Fontaine fables, rather than any remotely religious values.) These fables may not contain scientific information but they do establish the affinity of birds and humans.

The example of Oudry and Le Brun made clear that Buffon was not the first to be interested in what would become ethology. However much he reveals his background in Enlightenment humanism, he constantly endeavours to overcome mere anthropocentrism and to establish objective facts about birds. He is held back in this not only by his firmly held beliefs in Enlightenment values but also by his failure to follow Belon down the road of comparative anatomy in establishing firm definitions for individual species. His only instrument is observation.

The chapter starts with an extended account of the various classes of living creature in their use of the five senses. By adopting this approach, Buffon explicitly dismisses those capacities on the basis on which man bases his sense of superiority: 'ratiocination, discernment and judgment'. Buffon goes on:

> we shall consider only the different combinations of simple perception and endeavour to investigate the causes of that diversity of instinct, which, though infinitely varied in the immense number of species seems more constant, more uniform and more regular and less subject to caprice and error, than reason in the single species which boasts the possession of it.

The results of this approach are fascinating. It's little surprise to see in how many of the senses humans are way down the performance table: in fact, it's only in touch and taste that Buffon sees humans as scoring well. That sensitivity of taste has a regrettable origin, being a product of the effects of lazy opulence (i.e., being waited on). Buffon could hardly be more direct – taste comes at the price of 'the vices of luxury and opulence, indolence and debauchery'.

All Buffon's remarks are offered on the basis of his experience, observing nature in the field and on the basis of comparative anatomy, not between species within the same phylum (e.g., to establish particular bird species) and almost invariably between these phyla – for instance, comparing the anatomic structure of birds' eyes with that of humans or quadrupeds.

Buffon conducts his argument at two levels. He endeavours to show that the anatomical forms which give creatures their sensory pre-eminence affect their character and behaviour. So, the excellent sight of birds is associated with their speed of motion. Here the comparison is not with humans and horses but with the sloth; is it because the sloth has poor eyesight that he moves so slowly, or is his slow movement a reason why he did not develop better eyesight? Buffon doesn't say but he does argue for the effect of the physical make-up on animals' behaviour and character.

Buffon diagnoses in birds a marvellous ability to see the shape of the earth's surface and make of it 'a picture which exceeds the powers of our imagination' – something resembling but better than a map. The quadruped knows only 'the spot where it feeds, its valley, its mountain or its plain' and in consequence has 'no desire to push forward its exploration'. Birds think in pictures (that's a profound affinity with humans) and their flight is more than a physical capability, it's a sign of the breadth and openness of their mind.

If this gives rise to the view of the bird as a visionary and dreamer of distant and unknown lands, the other sense in which they excel – hearing – opens up intriguing reflections on bird song. The range of sounds which birds can hear and produce exactly matches that of humans. That's why we learn from 'finches and tits' how to sing, how to articulate our feelings. (Bats on the other hand have a language which humans cannot hear: no analogy offers itself here.)

Nowhere is Buffon's approach more in keeping with the contemporary state of science than in his extensive discussion of birds' sexuality. It brings him closer to Linné's classification of plants but unlike Darwin (with whom he shares not just the general interest in birds, but his fascination with bird song) he focuses almost entirely on the males. His views link in with sexuality, as he postulates 'a physical relation between the organs of generation and those of the voice'. The human analogy is of course between puberty and the male voice breaking.

Otherwise, Buffon emphasizes the huge role sexuality plays for the male birds. Caged birds 'have such an urgent need of coition that most sicken and die if that need cannot be satisfied'. He reflects on the enormous sex-drive of farmyard cockerels – though he feels sorry for them – human greed has reduced these birds 'to a kind of clockwork machine'. He discusses the physical form of copulation common to bird species, but it is left to his translator (in this case Oehme in a later volume) to outline the scientific basis of Buffon's view of reproduction. It had never occurred to Buffon to outline that theoretical background.

Buffon's problem is with the widespread sexual activity of birds, way beyond species boundaries. This is simply a given fact, a direct result of the males' strong sexual drives. For Blumenbach, in many respects a follower of Buffon, this fact makes a mockery of much human sentimentality. Even the dove, a symbol of faithfulness and fidelity, is, Blumenbach enjoys pointing out, highly promiscuous.

Buffon's indifference to the consequences of his observations is shown again when he fails to engage with a theory much debated at the time. Botanists and zoologists – and regrettably also eugenicists – were preoccupied with the issue of 'blood mixing'. Buffon observes a huge number of 'dreadful bastards' in nature. (Naumann, some 60 years later, was to be the only ornithologist to put such hybrids into his handbook and, as illustrator, to grace them with a portrait.) Buffon has no patience with those scientists who claim the sterility of hybrids as a sign of divine (or natural) providence. He also ignores the human analogies which fed into the racist thinking of the nineteenth century. He sees the cross breeds simply from the point of view of the science he is trying to establish. Hybridisation made it all but impossible to draw up an accurate inventory of separate species and to differentiate variations from genuinely new species. Cross breeding stood in the way of scientific ornithology. It's a refreshingly non-ideological attitude.

One major idea and one striking piece of imagination conclude our consideration of this chapter. The idea is one with which our age is preoccupied, though in ways of which Buffon could not have been aware – the influence of humans over nature and especially over birds. He starts from a reflection on the fact that the songs of bird species which inhabit the heavily populated parts of the globe are richer than of species inhabiting the deserts of Africa and America songs which 'have little claim to melody'. The richness of bird song derives from human contact – the finches and tits learn to sing from the humans. This leads him to a remarkably positive view of the Anthropocene:

> The real alteration which the human powers have produced on nature exceeds our fondest imagination – the whole face of the globe is changed, the milder animals are tamed and subdued and the more ferocious are repressed and extirpated. They imitate our manners, they adopt our sentiments and, under our tuition, their faculties expand.

No clearer observation of the dominance of mankind could be imagined, yet it would be wrong to see Buffon as welcoming every form of human intervention in nature – his comment on domestic cockerels made that clear – but the numerous analogies he sees between birds and humans are symptoms of his profound awareness that humans affect nature in everything they are, do and even say. The harmony in nature has a human origin, not to be gainsaid by observations that nature also influences humans. There is no awareness of human influence over nature in terms of the environment. The environment – in the text as in the illustrations – is taken for granted.

Buffon closes his chapter with a question which would have seriously embarrassed Victorian ornithologists. Having described the scarcely controllable sexual drive in wild and domestic species, Buffon reflects: 'Who knows what kind of love of this kind goes on in the forests?' Frisch's dream of the faithful and abstemious male is no more; take away the 'ratiocination, discernment and judgment' and the affinities of birds and humans take on unknown dimensions.

Individual Pictures

Buffon's work overflows with ideas and insights. Any reader will find many distractions before getting through to individual birds and this chapter has done the same, but now it is time to turn to individual species. We start with familiar subjects: the cassowary, two cranes and the bittern and conclude with three eagles.

Like Frisch, Buffon reports on the autopsies of the various birds carried out by the *Académie.* The cassowary is no exception and the result is an illustration of the digestive tract of the bird 'from nature'. Buffon, however, is not particularly interested in the details. For him 'it's much better to examine living nature than the entrails of dead animals'. Instead, he concentrates on the behaviour of this birds in captivity. He had every opportunity to study the birds closely, for he reports on their eating habits, the use to which they put their wings and their moulting cycle. He's particularly interested in the effect of captivity on the birds' aggression. Frequent human contact had considerably affected their character. Despite their wild and bizarre appearance, the birds in Versailles had learned to adapt to captivity and 'not been observed to be so strong or so mischievous'. Human contact had succeeded in 'meliorating their disposition'. Adaptation was a significant element in Darwin's theory, but for Buffon it was significant only as an indication of the harmony underlying the whole of nature and the affinity between humans and birds. In this case, Buffon's theory seems to base on the significant if banal fact that the birds had not killed anyone in the menagerie.

Buffon includes both cranes, the virgin crane under the species name 'king's bird'. The crests are called crowns. In Buffon's illustration, the cranes raise one foot, the emblematic posture we saw in Oudry's picture but the legendary background is of no interest. Buffon writes prosaically: 'When the bird is in a still posture, it rests on one foot.' As we saw, he tended to draw comparisons with dancers rather than with legends. His illustrators avoid giving the bird the kind of North European habitat chosen by Oudry – in fact there is no background at all.

While discussing the bittern illustration (**Plate 10**), Buffon has a few words about the technique of illustration and Oehme backs this up with comments of his own. Oehme criticizes Belon's picture of the bittern (notice how the whole tradition was known – illustrators and editors ranged freely across the centuries when looking for material). Oehme claims that it shows the bird in an uncharacteristic posture. Indeed he says, questionably, that it more resembles a kingfisher. In fact, Belon's bittern looks more like a duck, for it is portrayed on a horizontal rather than vertical axis. Buffon's picture – which Oehme takes over into his translation – is not unproblematic either. The bird has a strikingly flat head and the artist has stressed the thickness of its neck rather than having the bird raise its head and stretch and streamline the neck.

Plate 10: Unknown artist, *The Bittern: Ardea Stellaris*

Oehme discusses the possible reasons for these different postures. He ascribes them to the actual variety of alternative and yet characteristic postures the bird itself can adopt. (In other words, he does not see the posture as a question of artistic convention, which in many respects it was.) This range of postures 'has led to the variety of illustrations, which in turn have led people to conclude that there are so many sub-species of this bird'. His comments highlight the double function of illustration: first, the scientific purpose of an unambiguous identification of the species, preventing the proliferation of phantom species. But there is also what seems like an early hint of the modern field guide with its focus on characteristic postures (rather than detail) as the first point of reference. Behind both of these functions is the ambition to portray a 'typical' bittern. How, Oehme muses, can a bird adapt different poses and yet be characteristic in each of them? He raises the central issue of scientific illustration to convey the typical without details getting in the way. How can a bird be realistic in individual detail and yet represent the entirety of the species? We touched on this issue when considering Frisch and Dürer's little owls, and the problem is still worrying the translators.

Buffon discusses the bittern at some length. His interest in birds' character leads him to mention some of the anecdotal and mythical material which we discussed in connection with Rembrandt's bittern and it is in this context that he describes the bittern as a 'melancholic' bird. The majority of his material is to record the bird's behaviour in nature.

Two other features of this picture should be mentioned. First, that a shadow falling across the bittern's feet makes it appear that the bird has five toes. It's clear that the confusion has its origin in traditional aesthetic theory, which laid down where the light source (here the sun) should be integrated into a picture. The picture is correct, its information, however, confusing. For this reason, Schlegel devoted a section of his treatise of 1849 to detailed instructions on where the light source should be set in bird illustrations and on the avoidance of shadow.

The other striking feature of the picture is to exaggerate the plumage of the bittern, which – in keeping with its Latin name: *Ardea stellaris* – appears to be sprinkled with stars. In this respect, Rembrandt is more successful in reproducing the colours and pattern in the bittern's plumage because his hand is not led by any scientific purpose. Buffon's picture is another example of the scientific exaggeration of certain identification features.

Buffon discusses a further aspect of the bittern's behaviour, which other descriptions had left out, its pugnacity and its readiness to attack hunters and their dogs. Even birds of prey are reluctant to attack the bittern. This brings Buffon to discuss violence in the avian world. For once this is not yet another attack on the character of birds of prey, but concerns the preservation of the species. Buffon avoids the detail with which Frisch described the shrike but has to identify birds' daily struggle for existence. He has no way of reconciling this struggle with his vision of the underlying harmony of nature.

Oehme recognises Buffon's difficulty and feels obliged to explore this topic. He devotes a whole section to it, coming to the conclusion that this violence is part of 'the Creator's providential care'. This is a 'wise providence which serves to preserve a balance between the numberless creatures'. That thought leads Oehme to reflect on the relative numerical sizes of the various species. In a simplistic way he anticipates zoology's take-over of *Malthus' population idea

– a relationship which was an inspiration for Darwin's *Origin*. Oehme is not that far away from some themes of Social Darwinism, although vastly *avant la lettre*.

Plate 11: Frisch, *Black-brown Eagle*

The last pictures in this chapter are of eagles, those 'tyrants of the air'. Buffon's German readers may have felt some affinity with these birds because of their heraldic significance and this must certainly have been true of our first picture. The black-brown eagle, a picture which not only Buffon but both German translators took over from Frisch. (**Plate 11** reproduces the original but does not include the small additions which Buffon's illustrators introduced.) Buffon's picture did nothing to alter the dramatic qualities of the picture or the posture of the bird, just adding a touch of landscape – a patch of grass in the foreground and a prospect of the sea behind. These additions are similar to the background to the picture of the sea-hawk which Buffon took over from Seligmann's collection and which both Martini and Oehme included in their editions (**Plate 12**).

Plate 12: Seligmann, *Fish-hawk (Fischweihe)*

Both pictures are too static to be entirely convincing but both share a striking feature: the relationship between the eagles and their prey. This relationship features only in the illustrations: the accompanying texts make no mention of it. Before we turn to that relationship, however, it's worth noting where Buffon's actual interests lie. Above all, he challenges the idea that the 'black-brown eagle' is a separate species. Surely, he argues, there is no essential difference to the common brown eagle: 'the whole difference consists in the shades and distribution of the colour of the feathers, which is by no means sufficient to constitute two different species'. Buffon remains a convinced 'lumper'.

The two pictures are remarkably similar in the way they portray their subjects' prey. Disregarding small discrepancies in the proportionality – the rabbit seems large while the fish seems long and heavy – there's a more striking similarity. Neither the rabbit nor the fish show the slightest sign of the hunt. Both lie peacefully under their captors. The rabbit's gentle, mouse-like face is turned patiently upwards while Seligmann too shows his fish as untroubled, its eye and mouth unmoved. It makes no attempt to escape, neither prey writhes nor struggles. Both of them are uninjured, the birds' claws have not penetrated the skin and there is no blood.

Earlier we mentioned the acceptance of their fate displayed by the countless lions which lie dead at the feet of their killers in the trophy photos of big-game hunters. Even in death they seem to acknowledge the superiority of these humans. It would be an exaggeration to see similar consensus between victim and killer in these pictures, even if they look one another in the eye. But the scenes certainly put one in mind of the 'death conversation' in falconry, the moment when the victim accepts the distribution of roles between itself and the falcon. It's hard to know what is really going on in these pictures but we might see them as containing a gap, a hiatus, corresponding to an unresolved issue in Buffon's idea of nature. He is in no doubt about the character of the birds of prey: 'The birds of prey are more obdurate and ferocious than other birds. Accustomed continually to scenes of carnage and torn by angry passions, they contract a stern cruel disposition.' We see how this view bases on character, yet Buffon's problem to integrate this view with his understanding of the fundamental harmony of nature is never solved. The charm of these pictures comes from the incongruity in the relationship between victim and eagle, in Buffon's failure to square his idealism with the realities of the struggle for existence. Buffon was not the only one who could not square that circle, it was a problem for the whole of the period up to and including Darwin.

Our last eagle was not taken over by Buffon from another source. It is an American eagle but its picture is French, a product of his own team (**Plate 13**). Both the picture and the bird itself are dynamic but we are more struck by the bird's elegant posture and the picture's careful composition. In this, it is typical of much of his team's work. 40 years later, Audubon rejected Buffon's work and presented himself as a model of American exceptionalism. Yet this picture of Buffon's – though the landscape is arbitrary and suggests no relationship between bird and habitat – makes clear something Oudry's work also suggested, namely that Audubon took over significant elements from the French tradition, an elegance and sense of decoration which inspired his pictures too and added to their particular greatness.

Buffon's work lays claim to our attention still and modern reservations about the limitations of his knowledge and his methods will not disqualify that attention. The German translations of his work appeared at a time when German ornithology was in the process of formation. Buffon's European fame provided less well-known German illustrators with a wider public. We cannot use his work as a field guide, but if we want to understand the progress of ornithology and its illustrators, we should not ignore him. Behind all his writing is a love of birds, which has not lost its power to inspire.

Plate 13: Unknown artist, *American Eagle*

Chapter Six
A Dangerous Obsession:
Nineteenth-Century Classification

By 1800, bird books were reaching a kind of equilibrium as illustrations and print quality improved, while gastronomy and power faded as levers of consumption, others took their place and the competition between artists grew more intense. There was, however, a still more cut-throat forum for competition: classification.

It's easy to find these rivalries simply funny – for elements of them had an absurd quality and resembled medieval theologians debating how many angels could sit on a pin. It would be charitable, however, to see them as more than a bonfire of the vanities and to understand them as rites of passage of an emerging scientific discipline. Such a view doesn't necessarily make them more interesting. Whatever else, these debates overshadowed the practice of bird painting, and we must give an account of them.

We approach the topic from three angles. First, we give an account of the position of Lorenz Oken, one of the most well-known German 'life scientists' and a follower of some of the wilder strands of Romantic philosophy. Secondly, we look at the effect of some of these ideas on bird painting in Germany at the time. The complex classification principles, which Oken developed, crossed to England in the work of William Swainson and the final section looks briefly at the ideas which Oken encouraged in Swainson before examining how the dead hand of classification fell across Swainson's pictures. In a parenthesis during the chapter, I suggest the close parallels between Oken and Swainson's ideas and the work of chemists in establishing the periodic table. The unity of science can be observed even when science edges into the crazy.

Taxonomy – its modern definition includes nomenclature, classification and identification – had other casualties than Swainson's pictures. As soon as ornithologists understood that the endless proliferation of new species was not a threat to their profession (as Buffon had feared) but a once-off opportunity, there was no holding them back. The subject was so obsessed with discovering and then classifying new species that it could end up wiping out precisely the species they wished to classify. The pink Australian robin, *Petroica rodinogaster*, found its way into Gould's taxonomy only as a corpse – a situation reminiscent of the fate of the angels in theological dispute. Theories can be fatal not just to their proponents but also to their objects. The only advantage for the subject was that, because of the increasing involvement of generation studies (genetics before the subject could exist), when DNA classification came in, it could pick up on much preliminary work and move forward faster. In that sense birds – on the grounds of their relatively rapid lifecycle and clearly identifiable characteristics – were the *drosophila* of nineteenth-century genetics.

In another sense too classification could be fatal to its object. It was a heavy-handed implement with absolutist claims over nature. Linné's system aimed to subjugate the natural world. The totalitarian claim did not stop there. Because their ambition was to unlock the secret of the whole of nature, taxonomists soon got tempted into wider metaphysical speculations. At one moment they discuss the tiniest details of birds' beaks or toes, at the next the make-up of the universe and the purposes of creation. It makes for a strange reading experience.

The openness to sweeping metaphysics made the taxonomists' activity more dangerous as it got taken up by anthropologists or, as they called themselves in their worst incarnation, ethnologists. They transposed the authority of zoology into the classification of human beings. Like the ornithologists, they used a range of physical features, skull shape, nose, lips, angle of the face and last but not least 'colour'. The parallels did not stop there. They set up 'human zoos' in which specimens of the various 'races' could be gawped at by the European masses. At almost every development of the biological sciences ethnologists were at hand, turning tentative science into inhumanity. For every Darwin a Francis *Galton stands in the shadows.

Nineteenth-century classification took an infinity of different forms but had two principal directions: lumping and splitting. In the previous chapter we saw Buffon as a convinced lumper, uninterested in artificial categories, putting living things into larger, less divisive groups. He was the equivalent of philosophers such as *Herder and Locke, who accepted only one criterion for membership of the human species – upright gait – and ignored the kind of divisions ethnologists favoured. But Buffon's humanity was not modern, and in zoology and anthropology the splitters won the day.

Taxonomy Takes Over German Ornithology

One way to gauge this shift is to compare Frisch writing in the period before 1763 and Bechstein's handbook of 1791.

We saw how keen Frisch was to categorise his birds, how impatient he had become with a merely alphabetical inventory of birds. He managed to identify 12 classes but did not manage to find criteria for allocating birds to one or another class. As an honest man, Frisch starts his handbook with the list of the criteria he has tried and failed to establish as principles of classification. Habitat had seemed promising but Aristotle's classes of 'land air and water' birds didn't work because most birds belonged squarely in two categories. Early or late migration, the ability to imitate sound, feeding and washing habits – Frisch did not have the anatomical knowledge to use anything but superficialities (beak-shape, for instance) as criteria. For all that his work was a step forward in science, it was a failure of classification.

That Bechstein was in a different situation, had nothing to do with differences in character. He was no more systematic or scientific than Frisch – indeed, his starting point was field work, yet he was surrounded by systems. It would greatly have surprised Oehme and Martini to read his praise for Buffon's 'system'. His particular terminology comes from Merrem and his anatomical information from Blumenbach (whom Oken described as 'the worst professor I have come across in my whole life'). We can take from Oken's comment that Bechstein's science was not on the most solid foundation, but it's striking that within 30 years of the publication of Frisch's work, Bechstein is inheriting systems and doesn't have to improvise one. In his view there are various families, species and 'guilds' (*Zünfte*) – these are taken from Linné and Klein – then six orders taken over from Linné, augmented by three new orders from Blumenbach. The whole thing is rounded off by a reorganisation (what Bechstein rather naively calls 'the most up-to-date classification') taken over from Professor *Batsch in Jena. Bechstein has to admit that even Batsch's system contains 'some uncomfortable arrangements', but that seems to be a necessary part of all artificial systems'.

There were two problems with Bechstein's classification: how to identify classes and what to call them. One tendency was to move away from Linné's Latin. There had been voices demanding such a shift, in part as patriotic sentiment, grown stronger after the Napoleonic invasion of 1806. More serious voices looked to advance the popularisation of science by a change to the vernacular, but the linguistic nationalists got nowhere. Where the scientists, notably Oken, made a point of inventing new 'German' names for species and classes the result was a series of horrible neologisms or a reversion to dialect terms, both of which Schinz found 'barbaric'. An interesting comment from these debates, however, touched on one of the central issues of illustration. Names in the vernacular were effective in designating individual birds but lacked the power to abstract to a whole species – that was something Latin was good for. Bechstein encountered this problem too. If he moved away from Linné's Latin, he ended up with strange categories like 'half-swimmer' and even Naumann ended up with unmanageably long categories. Linné seemed simpler.

In ways like this classification was the cuckoo's egg in ornithology, driving out more real topics. It is typical that a series of books and dictionaries started to appear comparing various systems but having nothing to say about the birds themselves. Form had taken over from substance.

Bechstein's sources show a welcome modesty about their own systems and none has any sympathy with metaphysics. Merrem is typical of this attitude. 'For my part,' he writes, 'I think it more laudable to have collected material for a lasting scientific edifice than to have put together card houses which collapse at the slightest touch.' Batsch expressed the hope that 'the order which I suggest in the following pages and which is based on the observations of myself and of others will in time gradually approximate to nature.'

Merrem deserves better of historians. He not only discovered the anatomical principles of the air sacs in birds' respiratory system but he was an early historian of ornithology, being involved in the rediscovery of the thirteenth-century bird manuscript of Friedrich II. Merrem's own writings treated of broad topics and his publications on individual birds are just two in number, each dealing with five non-European birds. Here he focuses on identification issues, following a range of sources, mentioning: Linné, *Catesby, *Pallas, *Forster and *Dollmer. **Plate 14** is an example of his illustrations, *The rusty-margined flycatcher,* carried out by Eberlein and dated 1784. It's a strongly decorative picture, the birds are not explicitly carrying out a function, but at least the picture contains some specific information. The picture's symmetry is artistic rather than anatomical and though the birds are exotic (North American) the illustration does not exaggerate their exoticism.

Merrem highlights the range of expertise in German universities by the end of the century. His work is poised just before the boom in classification and for that reason quickly slipped into oblivion. System replaced sober empirical study. No one represents the wave which engulfed Merrem's generation better than Lorenz Oken (1779–1851).

Plate 14: Eberlein, *The rusty-margined flycatcher*

Oken

Oken is undoubtedly a major figure in the emergence of the biological sciences in Germany, but to approach him from the point of view of classification, as we do, hardly shows him from his best side. In fairness to him, therefore, I should briefly state four areas in which his achievements were long lasting. First, Oken founded the first national science association in Germany, combining the exchange of research and the advance of science. He also established a journal in which to discuss the national political and social issues involved in science. Secondly, Oken was a political progressive in an age where that was rare. One could argue that he was the first democrat to hold a chair at a German university. He exemplified, thirdly, the internationalism of his subject, maintaining close contacts with – among others – Robert *Owen, Geoffroy de *St-Hilaire, Georges *Cuvier and Louis *Agassiz. Finally, the world of theatre and opera appreciates that Oken supervised the doctoral thesis of the author of *Woyzeck*, Georg Büchner.

Germany in the nineteenth century was not really designed for this kind of figure and Oken's biography shows the deep scars which history inflicted on progressive figures. He saw little option but to go into exile in Switzerland, taking up the chair in Zurich. He passed through a brief period of hostility to France – an effect of the Napoleonic invasion – but the hospitality shown him by Cuvier during a research stay in Paris put an end to that (in view of what Cuvier, the great comparative anatomist, thought of speculative natural philosophy, his hospitality to Oken was remarkable).

A modern reader of Oken's works is liable to be confused by the mixture of themes and types of writing which they include. A huge energy is devoted to classification, some of it resembling the bed on which Procrustes welcomed his guests, stretching them brutally or cutting off bits so that they would 'fit': 'fudging' has always been an integral part of classification. He uses a wide range of categories, many of them for the first time – for instance, diet or nest behaviour – and his comments reveal both a determination to get the individual species to fit his categories and detailed specialist knowledge from field and laboratory. If his priorities are unclear, then it is because Oken's work is theological rather than zoological.

In fact, the word is teleological. The whole of nature is understood as a preparation for human beings and for the human faculties – an extreme case of arguments via 'final causes'. All living creatures are divided into four classes: feeling animals (also known as skin animals), sexual animals, intestinal animals and sense animals. The last class subdivides into four (touch being already covered in the first class): tongue animals (fish), nose animals (*Amphibia*), ear animals (birds) and eye animals (mammals). Here we notice faint echoes of Buffon's approach, but Oken reverses Buffon's emphasis on birds' power of sight. In arguing for these categories, Oken mixes speculative morphology with striking anatomical details, to which only a specialist would have access. A taste of these arguments: 'Is the insect not the human eye still unattached, the snail a still separated hand, the bird a human ear in the making?' In other words, in its classes nature is practising for the emergence of humans. These ideas mesh in with what Oken calls natural functions, for instance magnetism, which ends up bringing birds and horses together. It was not unusual to observe affinities between certain birds and animals: birds of prey and predators, for instance, or ducks and geese and ruminants but Oken seems happy to speculate. Yet we must remind ourselves that Oken held a professorial chair in respectable universities – that's not infallible proof of sanity, but he had a positive reputation with other chair holders and leaders of opinion and major scientists in the next generation, such as Rudolf Virchow, spoke respectfully of him. If his theories prove anything, it is that the opposition to 'materialistic' science ran deep and could cover a multitude of peculiar ideas.

Oken did not himself paint or keep a collection of birds. Perhaps that's just as well, given the odd opinions he put forward about them. The painters who could have done justice to his mystical ideas might have been Salvador Dali or Magritte. In a harmless way, elements of his thinking can be traced in the ideas of the old 'bird pastor', Christian Ludwig *Brehm, a kind of White of Selbourne in German ornithology. Typical of his work is **Plate 15**. This is both the frontispiece and the only coloured illustration in the volume and is the work of Johann Conrad *Susemihl, who was to illustrate Oken's empirical studies.

Plate 15: Susemihl, *Sylvia Wolfi*

Susemihl follows the painting convention of the bird sitting on a branch, not without justification since the birds belong to the *insessores* (perchers). There is a suspicion that the illustration was prepared from skins, but the point is that the birds are presented as belonging to different species. In his text, Brehm stresses the differential markers but refuses to recognise the two birds as principal species and variant. In that he shows himself to be a splitter but not without subtlety.

'Nature changes one basic form in the most varied ways. My *sylvia wolfii* seems to be a transformation of the Swedish blue-throat.' While that might sound like the description of a variant, Brehm insists that until we have a sense of the historical (he doesn't use the word evolutionary) priority of species, 'we must put all birds on the same level as species since we cannot determine which was the earliest.

Like many of his generation, Brehm was no friend of systems and he admits to knowing only two: Bechstein's and that of Meyer and Wolf (see Chapter Seven). He justifies this in a metaphor which has something of the mysticism of Oken: 'Living creatures radiate according to their different species in all directions, touching one another at the sides like the meshes of a net.' The metaphor, saying more or less what Swainson was to conclude, is a modernisation of the old *scala natura*. Brehm sounds humble in his acceptance of the limits of present knowledge but there is no expectation that if those limits widened, he would be prepared to give up the inner metaphysics of his views. He resembles the president of the University of Berlin in 1872, hiding his dislike and fear of Darwinism behind a gesture of humility – *ignoramus et ignorabimus* (we do not know and we will not know). It would be kinder to the president to relate his remark less to Prussia's banning of biology (that was a 'thou shalt not know') and rather more to Socrates' comment: 'The only true wisdom is in knowing you know nothing' either way Brehm's approach probably did less damage to the advance of science.

Nevertheless, this illustration and Brehm's attitude – similar to that of Oken – shows a clear link between splitting and illustration. The reverence for the species had two faces: the species touched the centre, proceeded directly from the hand of God. This belief explained the dislike of hybrids. Its other face was less idealistic. The more species, the more illustrations were needed. Splitting led – as Naumann and Schinz' correspondence makes clear – to countless mistaken identifications but it led also to countless commissions.

Oken's Successor: William Swainson and the Quinary System

If the poet Heinrich *Heine was right in his assertion that the English were a pragmatic people who (this is Oscar Wilde) were protected by their splendid physique from indulging in flights of the imagination, in contrast to the unpractical Germans dragging around with them, huge chunks of indigestible philosophy like stones in their bellies (this is *Nietzsche) – if Heine's view is remotely correct, then William Swainson is a serious exception to the rule. A talented bird painter, known above all in zoology as a proponent of the so-called 'quinary system', Swainson was a celebrated ornithologist, a member of various expeditions to far-flung parts of the globe. His name was mentioned in America and Britain in connection with Oken. His other sources were Robert *Chambers' *Vestiges*…and a little noted work, *Quinary System*, by Alexander MacLeay, a prominent member of the Linnéan Society in London. Swainson's writings display scorn for his predecessors, whom he describes as fit only to write out labels in a museum. The 'passion for collecting' had brought 'misfortune' upon zoology, yet in his ambitions, Swainson surpassed even Buffon, for he wanted his system lay bare the Creator's plans.

The quinary system views nature as divided into circles, each made up of five species from different phyla, such as birds, reptiles, mammals, fish and invertebrates, held together by affinities, which go back to the beginning of time. He takes over from Oken the idea that these affinities – or 'analogies' – can include relationships to inorganic matter, but most of his arguments concern living creatures. A particular species can belong to more than one group of five, so Swainson's works look like a series of Venn diagrams, overlapping circles, perhaps resembling a jig-saw with as many pieces as there are species in the world. Below this superior order, there are other groups of five (including birds) whose members are within the same phylum.

A feature of this jig-saw is that occasionally a piece is missing. When that happens, when a group has got only four or even three members, Swainson does not despair. The solutions familiar to the *aficionado* – look under the table, shake the box – seem to apply and Swainson is convinced that the missing piece will be found, it must be found either in as yet imperfectly explored parts of the world or in the fossil-record a creature will be found with the qualifications to fit the gap. That is an article of faith.

Two methods of Swainson's argument should be mentioned. The first is that the idea of an affinity between species from different phyla had been a familiar part of eighteenth-century ornithology. Not just Oken, Bechstein too wrote of such affinities and Linné too had seen affinities between ruminants and gallinae but these affinities were between groups otherwise defined. For Swainson the affinities were the fixed groups. Between a cockerel and the ungulates the affinity was the comb (or the antlers of the stag), for the spoonbill and the coati the affinity was that they rooted around in mud. There's also a whole sub-argument establishing groups through numbers – starting with five and then moving to the religiously significant three. This corresponds to Oken's confidence in 'the process of science becoming mathematical' but can hardly be taken seriously.

Swainson's work stands in the shadow of a remark of St. Paul that God is revealed in his works – 'For the invisible things of God are clearly seen, being perceived through the things that are made'. Swainson knows too much about the distribution of species across the globe to be persuaded by the literal truth of the Bible story of the Creation. The species did not all arrive in the Paradise Garden and then start spreading over the whole globe, but like Brehm before him, Swainson is happy enough to leave such matters on hold. It is unimportant, Swainson remarks with characteristic arrogance, 'for man to be informed at what era Australia began to contain kangaroos'. Science doesn't have to answer to all questions. That's a big difference to Oken, who wanted answers and cared about their implications for human society.

It's at this point with the quinary system looking shaky that I too want to indulge in an analogy and parallel Swainson's work with a scientific dream with a happier end. At the end of the excursus, we return to the birds.

Excursus: A brief history of the periodic table up to the end of the nineteenth century:

At the time that ornithologists were elaborating their complex systems in the search for the 'natural' order behind the innumerable bird species, a group of chemists were embarked on the same task in their discipline. Dalton, Béguyer, Döbereiner, Chancourtois and Newlands tried to bring order into their subject through classification. Like ornithology, chemistry was expanding fast and scientists felt the need for system both to structure their discipline and as an organising sense and pattern behind the world.

The chemists had the problem of determining which the elements actually were (the idea of an element included that form of permanence and purity which, until Darwin, was embodied in the idea of the species). When that problem was solved (and it seemed as far from solution as the ornithologists' attempt to classify while new continents were being opened up to their science), they had the ambition of understanding, of seeing an order, a plan even, in the total assemblage of elements. Of 114 elements known today, Dalton knew only 43. By the end of the century that number had gone up to 63 and on that basis the great Mendelejev set up the basis of the table we know today. Though they were aware of the fragmentary nature of their knowledge, these chemists worked determinedly on a system of all the elements, not just on the smaller internal groups (for instance, the chlorine, bromine and iodine triads). Their goal was nothing short of the total picture: to discover the fundamental order of the universe. Their search lasted a century.

Döbereiner went on to identify further triads of elements with identical properties, but he and his colleagues had serious difficulties bringing the other elements into order. Like all systematisers they faced the problem of an irreducible residue, those elements left over and not fitting in anywhere. Döbereiner's work further illustrated the problem of classifying objects by means of deficits, by means of what they are not or what features they do not possess. At the same time Newlands and Chancourtois came up with the law of octaves (later known as the law of periodicity). This established that every eighth element had features repeated in that series – a fine idea, parallel to Swainson's numerical searches, but it left the problem of the other seven which were not in the row. Some of the classification criteria involved significant features of the elements – one could say that these criteria ultimately led the way to the discovery of the role of atomic numbers – while others involved the elements' ability to react with other elements; further groups involved more peripheral features, such as colour and smell (i.e., anthropocentric).

The chemists working on this problem were modern in their approach, relying on the laboratory rather than on idealistic philosophy. Nevertheless, there were aspects of their thinking which went back to seventeenth-century scholastic philosophers, for instance to the understanding of the four-fold forms of affinity and the idea that every element relates to an ultimate essence of things and possessed its own 'signature'.

There was a still closer affinity between chemistry and zoology. Both systems had a strong predictive element. The chemists regarded any gap in their charts (the missing jigsaw pieces) as indicating an element which had not yet been discovered. The validity of Mendelejev's system was established partly by its success in predicting the existence of the elements germanium, gallium and scandium. The final result of these researchers – the periodic table as we know it today – is the product of Mendelejev's discoveries but its real foundation is a system which was not available to Mendelejev – the sequence of atomic numbers. Zoology cannot compete with such completeness, even if DNA identification offers zoology comparable objectivity, its results do not smooth over the lack of mathematical tidiness in nature – what Darwin called nature's 'higgledy-piggledy'. Nor are the species as permanent as elements, even before the Anthropocene started their whole scale destruction. But the periodic system remains an important dream of the eighteenth and nineteenth centuries, fulfilled in the present: a dream from which zoology woke up disappointed.

Swainson (cont.)

I want to raise two brief issues before returning to our central topic, illustrations. The first is that Swainson absolutely refused to admit any developmental principle into his vision of species. Even in Oken's odder moments – when ears fly around in pre-development stage of humanity – there was an acceptance of the idea that mankind had developed. Swainson rules that out. His vision is of a static natural world, fixed for all time in its circles (Darwin called them 'vicious circles'). Swainson never forgave Cuvier for basing his research on the idea that man was an animal. For Swainson man had always been the high point of creation.

Swainson moves on to more interesting territory when his expresses his view of species – the basic unit of bird painting. While Swainson's quinary groups represent a significant and often stupid devaluing of the species – 'In nature groups are more concrete than species,' he writes – he does have a vision of species emanating from the centre, just as in Brehm's image: issuing from the hand of God. Species are

> an element in nature, whose deviation from the elements which surround it is – as far as we can observe – permanent, so that we can assume that it experienced its identity in the first moment of its existence, in the moment when it issued from the hand of the Creator.

Ignoring hybridisation and closing his eyes to the evolutionary origin of species, Swainson does offer here a clue as to the inspirational basis of bird painting – the sense of the uniqueness and specialness of each individual species.

This sense brings Swainson back into the orbit of all bird watchers and painters. Perhaps it was this sense which made people think of birds as symbols of the individual soul. Darwin knew that feeling too – his excitement at the manifold life teeming on his 'entangled bank' is similar, if secular in tone. Häckel and Jacobsen, sitting over their microscopes (and in Häckel's case drawing the most beautiful pictures of his radiolarians), shouting out *vivant cellulae, vivat microscopia* – all these experiences have the quality of a revelation, channelled through scientific observation rather than prayer, but the sense of excitement at the natural world is the same and it is the basis of painting.

There's so much religion in Swainson's work that it seems unlikely that he could ever paint a bird naturalistically. We might expect him to paint the pelican among the risen saints or at least the lonely bittern in its stagnant pool. But the greatest charm of bird painting is that even the most ideological are persuaded by the birds themselves, taken in by the colours and forms which they offer. So we should not give up hope.

Swainson's pictures

Swainson was well aware of the importance of the pictures in the bird books of his day, indeed they had created a luxury market too expensive for many readers and libraries. Pictures therefore failed to meet their 'general usefulness' – in fact he obviously feels disadvantaged as reader. That's one reason why he launches into attacks on many of his predecessors, especially Buffon, Audubon and Temminck.

In his *Zoological Illustrations*, a miscellany of birds, butterflies and insects, Swainson's science comes back to intrude on his painting. No more do we hear of the inspirational effect of recognising that species come directly from the hand of the creator. Instead, we read that in the illustrations the task of the artist is 'to round off the details found in scientific monographs'. Science dictates and art merely 'completes'. Far from praising, as once he did, birds' 'elegance of form, beauty of colouring or sweetness of voice', Swainson focuses on the 'elevated views of the philosophical zoologist', for whom the study of nature no longer consists in 'the study of words, the retention of names, or even the accurate description of species'. 'Philosophical' zoology has broken painting's link with nature, science has lost interest in its objects.

Unfortunately, Swainson's claim 'that the system is more concrete than the individual species' is born out in his picture (**Plate 16**), the *ramphastos vitellinus*. Here Swainson is ticking off another piece of his jigsaw; despite his extensive travels in Brazil he had never actually seen a toucan. In contrast to Oudry's toucan, Swainson's lacks style or exoticism. Despite criticising the routines of bird painting, Swainson follows the oldest of clichés and puts his toucan on a branch, where – with no exotic or mysterious qualities – it looks like a boy forced to pose for a school photo.

Plate 16: Swainson, *ramphastos vitellinus*

We end with one of Swainson's remarks on Audubon, which tells us something about the market for bird illustrations. He claims that Audubon's pictures 'have not been made use of for generalisation'. This can mean only that Audubon had not become one of the canonical illustrators whose works were tapped every time a new anthology or handbook was required. That's true (and money will have played a part in this) but Audubon had left a monument to the energy and beauty of North American birds and had revealed them to the world. Classification could not lay claim to any such achievement and Swainson's system had become an end in itself, 'science' had reduced birds to a skin. The system had become more concrete than the individual species. It was time for bird painters to get back to their first inspiration: the individual birds.

Chapter Seven
A Peasant Farmer Goes to School

It's with some relief that we leave the classification debates with their high-flown theories. Not many of them touched what fascinates us in birds and what interests us more – pictures of living birds – was not richly rewarded either. From the less than dramatic images with which we had to deal, we turn now to one of the most significant figures in the development of scientific German ornithology. Oken and Swainson saw ornithology as a fruitful topic for theory and speculation, their debates being carried on far from muddy fields and hedgerows but our next subject could hardly be more different. Leaving school at 14, Johann Friedrich Naumann (1780–1857) found himself taking over not merely the few wet acres which his father had farmed before him but also the legacy of his ornithology. It was the fact that the fields regularly flooded and could not be worked which left them time to watch the many birds attracted onto their land. Johann Friedrich then overtook his father's achievements, finally obtaining a doctoral degree in zoology. His artist's studio doubled up as scientific laboratory.

Naumann was not the only prominent naturalist to be marshalled into the activity by their father. The author Alfred Brehm was pushed no more ceremoniously into the line by his domineering father, Christian Ludwig Brehm, the so-called 'bird pastor'. We note, incidentally, that Christian Ludwig's efforts to force his son into his own beliefs and profession failed conspicuously; indeed, Darwin was later to spot Alfred Brehm as a figure anti-clerical enough perhaps to help him in his own scientific projects. He did become the author of the century's most abidingly popular book of zoology, always known as *Brehms Tierleben* (Animal Lives). Naumann too appears to have been happy to go in the direction he was pushed.

His father, Johann Andreas, had got as far as he could in ornithology. He had produced an extensive practical guide to setting traps for birds – this skill involved significant information – and had written a short inventory of German birds. Within this project, the old man's first requirement was for a proper illustrator in the family, so at 14 Johann Friedrich was sent out to get the necessary skills, starting with engraving – with great success. However, Johann Friedrich did not stop there. He could see that ornithology was going nowhere without a much stronger basis in science and it says much for his ability and for the accessibility of universities that, initially self-taught, he embarked on a systematic study of scientific ornithology. He got his books from the natural history society in Halle, some 25 miles from his wet fields at Ziebigk. Subsequently he struck up a friendship with Christian Ludwig *Nitzsch, professor of natural history at the university of Halle. Under Nitzsch's guidance, Naumann learnt scientific anatomy. Within a few years Naumann was contributing scientific articles to the specialist journals and by 1840 he had obtained his doctorate.

This was an impressive rise but Naumann's background stuck with him all his life. While Oudry moved in the atmosphere of Versailles and while Oken and Goethe squabbled over books in Weimar, Naumann felt uncomfortable in most social milieux. Throughout his life we see him struggling financially, trying to find cheaper ways of sending bird skins to his correspondents, always buying the cheap edition of any new handbook and then colouring the illustrations himself. He hardly moved away from his patch and had to be especially respectful to the local aristocracy in order to watch birds in the castle park in Köthen. (It was in that castle that 150 years before, Bach's *Brandenburg Concertos* had received their first performance.) As he became better known, he corresponded with fellow ornithologists across Northern Europe and exchanged information, skins, eggs and spare copies of books and articles. That's why postage costs were so significant for him.

This may sound quite modern, the kind of sharing which has always marked bird watchers. Yet there were a number of inhibiting factors which got in the way of academic exchange. These came down to the splintered and uncoordinated state of Germany in his day. The post was not reliable (almost every parcel and letter he sent was in effect 'international' since headed to a different state). Despite Oken's innovation in founding a nation-wide scientific society with a corresponding journal, scientific work was unevenly spread across the country.

In particular the kind of popular, progressively scientific journals which were to do so much to modernise Germany – we think above all of *Die Gartenlaube* and *Die Natur* – were big city institutions, miles away and out of reach to Naumann. He could not afford extensive travel (he did not undertake any field-trip outside Germany before 1835) and even travel to important ornithological habitats, such as the sea or the mountains, was expensive and complicated by a proliferation of independent states and a plethora of internal frontiers. Naumann was pinned down on a small patch of land and inhibited in extending his circle. For some years he maintained a correspondence with Schinz, professor of zoology in Zurich, but, to Schinz's irritation, continued to address him as 'your honour'. The habits of Swiss democracy had not reached Ziebigk.

There are few better indicators of the onward march of ornithology than to compare Naumann senior's handbook (1791f) with the standard, 12-volume handbook of his son, which started to appear in 1822, initially as a re-edition of his father's work. No one was more conscious of this than Johann Friedrich. These days, he wrote in his preface, it was impossible to write about birds without an overall 'theory of science'. However, much he respected his father's labours – 'a countryman grown grey-haired in the capture and hunting of birds' – his father's pragmatic feeling for 'realism and truth' could be continued only by expressing all observations in 'an ordered system in keeping with the best science of the day'.

Johann Friedrich's scientifically modernised version of his father's work starts with an extended introduction into the anatomy of birds, drawing not only on the work of his friend Nitzsch, but also on Cuvier. The subject's internationalism asserted itself, even in the German provinces. Naumann explains in detail the effect of microscopy on ornithology and mentions as an example the insights which it offers into the colour of feathers – a topic of abiding scientific interest. While this had been unproblematic to his father, Naumann points out that the feathers of a green woodpecker were in fact yellow. His discussions of the relationship between intelligence and instinct – his terms are 'cleverness' (*Klugheit*) and 'art instinct' (*Kunsttrieb*) – is carried on in the exact language used by Schinz in his own handbooks. There is an accompanying analysis of bird brains in dissection.

The second chapter concerns the 'outer life' of birds and here Naumann discusses a wide range of topics. Some of these are old-fashioned, such as the 'usefulness' or the 'huntability' of various species and it's not surprising that there are also sections entitled 'trapping and hunting' but the overall impression remains modern and scientific.

Specifically, Naumann vehemently rejects the idea of birds hibernating in a marsh. He bequeathed to modern ornithology the term *Zugunruhe* to designate the nervousness of birds before migration. He has interesting comments on birds' imagination – a product of observing birds when they are asleep (again a symptom of more archaic methods of research). On sexual matters Naumann is rather more reserved than Buffon had been, conceding that birds show 'striking examples of immorality' but he is less worried about the effects of hybridisation. Indeed, he uses the full page picture of a *Bastard of the grouse and the capercaillie* **(Plate 17)** as the frontispiece to the sixth volume of his work. This is a striking choice in view of his contemporaries' obsession with 'pure' new species. Like Buffon, he is sentimental about domesticated birds, discussing their 'love of freedom'. He is a determined opponent of hunting for pleasure and as a sport but he has no embarrassment in explaining to Schinz that in pursuit of his scientific research he shoots between two and three hundred specimens a year.

Plate 17: Bastard of the grouse and the captercaillie

A great deal of Naumann's time and energy went on classification, primarily in order properly to identify doubtful skins or to correct 'the endless new species which Brehm dreams up'. He is critical of those ornithologists who 'attach too much importance to insignificant features' of a species. His own dream would have been to find an entirely natural system, which avoids 'putting birds into separate groups, of which nature would not approve'. In terminology he stays close to Illiger, although it was Naumann who established the word for the front part of a bird as *Brust* (Illiger had wanted *Bauch*, stomach). As he revised his father's work, he cleared out numerous archaic and dialect names for the various species. His father had, like Buffon, enjoyed the variety of names more than he disliked the potential confusion of species.

Forms of Identification in Other Disciplines

Through the historical sequence of bird books, the meaning of the word identification changes radically. The earliest bird books presented knowledge about birds simply as part of general knowledge. They were one of many variants on the illustrated lexicon. That's why Gesner had no problem including illustrations of a phoenix, alongside pictures of real birds and describing them in similar terms. He wasn't intending his book to help his readers identify a phoenix – the item was included, like all the others, as part of general information.

By contrast, Baldner's inventory of regional birds was based entirely on personal observation and practical experience and in intention his text shifted away from a lexicon as identification in the modern sense gradually came to play a role. Frisch followed this example a hundred years later, albeit focusing on a more scientific form of identification, indeed to a large measure *post mortem*.

Buffon and Bechstein attached especial importance to observation in the field and Naumann, despite his systematic training in anatomy, shared that focus. His work was like a multiple volume field-guide, providing in picture and text an initial means of identification and a more scientific follow-up. Observation in the field could be confirmed or corrected on scientific authority, while observation was important enough to be allowed to revise, to add to, or to question scientific statements. As Darwin commented in a letter 'How odd it is that anyone should not see that all observation must be for or against some view if it is to be of any service' (a remark relevant to illustrations). Naumann's text combined experience with science. In this his work was typical of all the emergent sciences in its ability to integrate – as a historian of science put it – the 'local knowledge of lay persons' with innovatory research.

Meanwhile the sciences were not just going their own way, there was cross-fertilisation between them. In particular, the emergence of the social sciences depended on that interplay of personal experience and science, of individual case and general rule, which marked Naumann's work. They established their legitimacy not just by adding new disciplines – for instance, ethnology or sociology – but also by exporting their insights and methods into other fields. As we shall see, Naumann's work draws attention to striking parallels between ornithological field-guides and the mid-nineteenth-century novel.

Painters and ornithologists were hardly more conscious of what was happening in other disciplines than were the scientists – people tend not to look outside the field they are trying to succeed in. Frisch, for instance, saw himself contributing to science but not to the Enlightenment. Nevertheless, some scientists through contacts with writers and intellectuals followed in general cultural trends and benefitted from these cross-disciplinary contacts. Darwin, surrounded in his profession by scientists who looked only to the church for extra-scientific ideas, made his breakthrough by becoming aware of the contribution which Malthus had made in the social sciences and incorporating it into his own studies.

Germany offers two striking examples of this cross-fertilisation. Sometime after Naumann's death, a remarkable congruence of ideas could be identified between the biological and social sciences. The sociologist Ferdinand *Tönnies developed an analytical instrument labelled with the contrasting concepts of 'community and society' (*Gemeinschaft* and *Gesellschaft*). On the basis of these concepts some of the most significant analyses of industrial and urban society were carried out. There is a clear link between Tönnies' work and the groundbreaking studies of K.A. *Möbius in biology – work which culminated in the concept of the 'life community' (*Lebensgemeinschaft*). This term expressed the mutual belonging and interdependence of varying organisms within the biotope. Möbius and Tönnies identified both in nature and in society the existence of powerful forces which could be described as community or home (*Heimat*). Indeed, Lynn K. Nyhart has called the biology of that time 'a discourse through which nature and society could be discussed'. The influence of Möbius' and Tönnies' ideas on social and political thought was enormous and in themselves ideologically value-free. Yet *Heimat* can be instrumentalised to emphasize particularity, specialness, exclusivity and ethnicity (i.e., racism), just as community can be presented to stress the diversity and interdependence essential to successful groups of organisms, so to speak their multiculturalism, what Darwin affectionately called the 'higgledy-piggledy' in nature.

In Chapter Ten we examine the relationship between the work of the Swedish nature painter, Bruno *Liljefors and the particular form of Darwinism popularised in Northern Europe by the Danish biologist and novelist Jens Peter Jacobsen. The comparison makes sense on the basis of the close relations of artists and intellectuals to science in the Scandinavian countries. For Naumann stuck in the German provinces, no such overview was possible. Yet there are striking parallels between his work and the forms of social observation behind the work of one of the world's greatest novelists.

These parallels are most clear when we look at Honoré de Balzac's series of 'physiological sketches', which appeared from 1840 and in which Balzac set out to identify the significant social types of his society. For the social scientist, the social type is what the species is for the ornithologist: the building block of society. Species and social type share another feature. Both include a huge number and variety of examples – bird species considerably outnumber mammals, for instance – while social types can be endlessly identified in professions, classes, ideologies or even in the Aristotelian humours such as melancholic or choleric. We take as our model text Balzac's sketch of the *Épicier*, the Grocer.

The text describes the grocer, his clothing, his intellectual world and his social function. The illustration was absolutely crucial to the genre, it was entrusted to the celebrated illustrator Sulpice (Paul) Garvani and clearly copied from field guides. It shows the grocer in a position of subservience to his customer – the 'characteristic pose' for which the bird painters aimed. The text describes the shop (the habitat), the goods on sale (indicating the nature of the biotope on which the grocer draws) and the social customs of which these goods form part (what people have for breakfast, for instance or the levels of luxury on which the grocer's clientele exist).

This approach was taken over as a standard technique of the entire realist novel in the second half of the nineteenth century. While the heroes of eighteenth-century literature had been unusual people, exceptional even (in describing early bird books we called such types 'exotic'), the nineteenth century watched its literary heroes become ordinary, members of their society, social and historical *types*. The huge range of the novels' characters (not just the eagles: smaller, everyday species became heroes) showed typical features of society and their interaction represented the interdependence of diverse social forces. Their occupations, dress, houses, the colours with which they surround themselves were not shown as reflections of individual taste – all those things were shown, as in the *Grocer*, to be expressions of general social values and functions.

The idea of the socially typical reaches into the characters' physical body. The grocer's social environment determines his face. The curvature of his spine, his gestures: all this is explained – in short, his 'body language'. Indeed, this metaphor itself goes back to the early nineteenth century, as social scientists attempted to express the task they had set themselves. They compared their science to learning a new language: human behaviour had its alphabet, its individual words and its grammar. And because there was no other way to read that language than with the eyes, illustrations were essential. In striving for that understanding the novelists were strongly influenced by two fashionable sciences of the time, both poised between biology and social knowledge: physiognomy and phrenology, both claiming to read the body, not just as anatomy but as a key to society. Readers got to know what ginger hair or sensitive fingers might mean. The novel became the setting for a practical comparative anatomy, a field-guide to the species in human society.

It was therefore no accident that Balzac thought of his work as a 'zoological novel', for he took over the scientific forms of the discipline. Earlier we discussed the unfortunate parallels between zoological classification and the racism to which it lent a scientific patina. Ethnology divided up the nations of the world with something of the self-righteous callousness with which the early ornithologists went about their work. Many Europeans could claim, with Baldner, that they had shot all their specimens. It wasn't just 'popular' literature which stood in the shadows of contemporary zoology: 'high-brow' literature (the ghastly phrase itself a shake down from *Camper's racist classification) shared a visual approach to identification, making questionable rules and norms from experience, testing out these rules on the unresisting objects of their gaze. While the hierarchical model of classification – Buffon's preference for eagles, Oudry's painted royal menagerie – had given way to the model of mutually dependent organisms, the value judgments behind the models did not change. And these views were fixed on a dissemination model which paralleled text and picture around the identification of species and which equally faithfully served ornithology, the novel and social understanding.

So, probably without having read a word of any novel, Naumann had not only stabilised the discipline of ornithology: he had seen its approach established in the mainstream of the progressive literature of his time.

Science and Art in Bird Painting

With the stabilising of the format of the handbook and the increasing competition between the bird painters, it was to be expected that debates on the two variants of bird illustration would restart. In England Swainson provoked the debate with his discussion of the contrast between the artist's pencil and the scientist's description. Later in the century, Ruskin continued to treat science as the opposite, indeed the enemy of art. It was in 1849 that Hermann Schlegel tried to bring some order into the debate in a study *On the Purpose and Properties of Natural History Illustrations*.

Schlegel's qualifications for writing this essay – despite being born in Germany, he had had the benefit of living in Holland – were typical of ornithologists in a colonial power: he had published studies of the avifauna of the Netherlands, of the East Indies and of Madagascar. He had produced a standard work on falcons – Josef Wolf started his career with contributions to that volume – and the breadth of his experience underlines the confinement in which Naumann had to pursue his own ornithology.

Schlegel's thinking starts from an absolute distinction between the art and science requirements of an illustration. The fact that he saw these as polar opposites says something about the emergence of a technically understood science of ornithology.

> The scientist wants the illustration to include the last detail. He cannot see why one and the same colour can have so many nuances as it moves from light into shade and he expects the artist to achieve things which simply cannot be done or which would inevitably make the picture artificial and offend all the rules of art. For his part, the artist will be unfamiliar not only with the object itself but also with the purpose of his drawing. He wants to see everything in a picturesque manner and has no interest in the details, whose scientific importance he cannot grasp.

Schlegel's starting point not only shows how far science has moved but he disregards a whole series of things with which we have become familiar. He doesn't take into account the idealism of Ferdinand Frisch or the possibility of double talents, such as Naumann and Swainson. Schlegel is focused on the division of labour which characterised the relationship between Darwin and Gould (his 'invisible helper'), the separation between field work and the archiving of the material. Schlegel also leaves out what Humboldt regards as the naturalness of a double talent, the communality of understanding between an aesthetic and a taxonomic eye. Such a mediation between science and art doesn't occur to Schlegel, although Darwin's advocate in Germany, Ernst Häckel, reminds us how in the days before sophisticated photography the hand was the only link between the microscope and the outside world.

Schlegel has direct advice for the artist as to how he should go about creating pictures in either category. The artistic picture would benefit from uneven lighting, a lack of symmetry and a certain disorder in the bird's plumage. The scientific portrait would give no priority to any individual feature of the subject or any part of its anatomy. It would have to be precise about the number of the various types of feather. It was for this feature of his work that Schlegel singled out Susemihl for particular praise, though strangely enough Schlegel's preferred bird painter, Josef Wolf, dismissed Schlegel's feelings on this point. He claimed that the range of feathers was so great that no artist would perfectly portray them all.

None of this was particularly innovative but where Schlegel's essay had a clear usefulness was in summarising the dilemma we encountered in earlier artists. The illustration had to portray what was typical rather than individual in the bird. This related not just to the 'characteristic pose' which had always been required but to the actual appearance of the bird itself. The scientific illustrator, according to Schlegel, had 'to leave out the individual features of the object for the depiction of the individual in science has to represent the whole species'. In making this demand, Schlegel does not seem to realise that he has just blurred any basic differentiation between scientific and artistic illustration. He also shows less awareness of the conventionalities of both forms and their profound affinity in that – in Salomon Gessner's words – both forms have to give the user 'the illusion that he has the object itself in front of him'.

In some ways the real force separating out art and science in illustrations came less from the practice rather than from idealist aesthetic theories. Classical aesthetics always felt a strong compulsion to distinguish surface realism from any deeper apprehension of the object. It's one thing to insist that art 'transfigures' everyday reality, that it aims for 'truth' rather than 'reality' but these are hardly useful categories in connection with bird paintings. By the time the artist has cut out the mythological and ideological understanding of nature, so that the little owl is a little owl and nothing else, it's hard to know where truth has gone. In any case, an interpretative element is present in scientific illustration; a proper bird painter has to understand what he sees in order to see the function of features as well as their appearance. In an article written a hundred years after Schlegel, an entomologist explained that 'every scientific illustration should include as far as possible a result, not just "the appearance, the recognisable and the visible" but no less "form, understanding and interpretation".' These are aesthetic and scientific accomplishments.

The debates which Schlegel aired recur, not just among historians and theoreticians but among illustrators. Debates about the contrast between art and realism accompanied the emergence of nature photography and its claim to surpass art. Schlegel's problems have not been solved in any case private and public acceptance of what art is has stretched beyond anything he might have imagined (any discussion of Campbell soup cans as art would have been alien to him). But, as befitted a senior official in a prestigious museum Schlegel saw no reason why he should hold back from passing judgment on the bird artists of his day. He is hard on Audubon – suggesting he is not good on the bird's character and that the quality of the engraving is often heavy-handed, criticizes Knip for the quality of the feathers she paints and is faintly damning of Naumann by referring to him as 'much more the naturalist than the artist'.

The Pictures

Schlegel's criticism of Naumann's pictures, like so much else in his essay, is not susceptible of proof but it seems helpful to look at his pictures within the competitive context that Schlegel's essay suggests. The obvious point of comparison is a major work of which the first two volumes had appeared in 1805 and 1818 before the first of his 12 volumes came out (It's worth remembering that 1805 was the year before Napoleon's military victories, invading Germany and driving the Prussian monarchy into exile and humiliating all the German states. 1818 was only five years after the battle of Leipzig began the process of driving Napoleon out of Germany and only three after the Congress of Vienna had established some provisional and unsatisfactory order in the German states. These were not uneventful years when ornithologists could follow their science without distraction).

Its authors, Bernhard Meyer and Johann Wolf, belonged to a different class to Naumann. Meyer was a doctor and pharmacist, the principal teacher at an industrial college and, like so many professionals at this time, a member of a natural history society. He met Goethe a couple of times and found his way into Goethe's biographies. He had held personal conversations with Alexander von Humboldt. This was, in short, a person with whom Naumann could not compete: the question was whether his book could.

Meyer and Wolf's book is printed in folio size with more than 120 full-page illustrations. The coloured engravings are made from pictures by Ambrosius Gabler, engraved by J.C. Bock with an occasional picture by J.M. Hergenroeder. In the preface Meyer and Wolf stress the quality of the illustrations as their book's chief selling point. It is broadly similar in organisation to Naumann's but the sequence of birds is more arbitrary. The authors point out that the illustrated pages are not numbered, so that it is possible to jettison the text and create a picture book. Naumann was more modest about his own illustrations but the obvious advantage his work had over its rival was that – while Meyer and Wolf seemed ready to have their text discarded – Naumann started with pure science: anatomy, a skill which he had built into his studio. It was this commitment to science which gave Naumann's volumes the edge and indicated that there was no longer a public for Meyer and Wolf's work. If we take two eagle-owls, it might show us the relation between the illustrations in their two works.

Naumann's picture (**Plate 18**) is a powerful portrait, intensified by the lack of natural background. We can see that a rudimentary background has been indicated but not fully executed and we suspect that, had Naumann had the money, he might have got someone else to paint the background (as Audubon did on occasion). The ear feathers stand out, so do the claws. The pattern of the breast and wings is carefully and clearly recorded, perhaps the eyes are a little large. The wing feathers are particularly distinct. The lighting is even, but the picture avoids symmetry. Naumann identifies the gender and age of his subject, having discussed in his text the variations in marking which depend on these factors. He shares with Buffon the recognition of the need to have four specimens of a species, plus one nestling in order to record it properly. He is as far as I can see unique in German bird illustration at this time in recording such matters with his illustration.

Plate 18: Naumann, *Eagle Owl*

The equivalent picture is Gabler's *Eagle Owl* (**Plate 19**). When I first saw this, I was really excited. The copy in the British Library appeared not to have been opened for well over a century, a little of the paint had come off onto the opposite page and the dramatic effect of this, the opening picture of the first volume was considerable. Yet, as one looks at the picture for longer, it comes to feel familiar. The stylised piece of ground on which it is standing – this a standard feature of most of Gabler's pictures, as in his picture of the shrike (**Plate 19**) – the emphasis on the aesthetic rounding of the picture: all these features seem on reflection artificial.

Plate 19: Gabler: *Eagle Owl*

Part of the drama of the picture comes from the fact that the bird has not simply found its prey but has killed it and torn out some of its innards (Naumann also has pictures of birds with their prey, by chance his eagle owl is not one of them). Perhaps there is something artificial about the innards themselves and the way in which the bird's claws are resting on them rather than penetrating them. Among the decorative effects one might mention an excessive regularity in the markings an impressionistic account of the wings – certainly of the wing-feathers. In portraying the ear feathers, Gabler seems to follow Schlegel's recommendation in introducing a slight disorder. But it is indeed a fine portrait.

We could say that both portraits mark a new naturalism in bird illustration, although the description of Gabler's work as a new beginning 'after the end of natural history painting' seems more appropriate to Naumann's portrait. Naumann develops the scientific tradition of the eighteenth century while Gabler holds on to the aesthetic side. Gabler, with his softer focus and correspondingly softer brush-work, certainly does not foreground the scientific purpose of his picture in the way Naumann does. In many ways, despite their ostensible identity of purpose, these two pictures could have been the basis for many of Schlegel's comments. The failure of Meyer and Wolf to continue their series implies that their texts, together with Gabler's illustrations, did not meet the expectations of a scientific age. But we lack criteria to say that Naumann's picture disqualifies itself as art.

Plate 20: Gabler: *Shrike*

Naumann achieved much from staring over his flooded fields. He showed that the door leading to science was open to the diligent. He consolidated a writing form in ornithology which became a model for great novelists. His book marked the end of courtly ornithology and sharpened awareness of the illustrator's task. Yet in all this he stayed close to the practical, 'countryman's' skills and pleasures of his subject and made possible the parallel development of the delights and the scientific rigours of ornithology.

Chapter Eight
Josef Wolf Takes London by Storm

With Josef Wolf (1820–1899) we reach a highpoint of German bird painting. It's not just the dramatic quality of his pictures and their technical brilliance which draw our attention but the importance of his subjects and the distinction of those who commissioned his work. We will look at his work for Darwin but note in passing the commissions he undertook for the evolutionary biologist Alfred Russel *Wallace and for the Africa explorer David Livingstone. It was his work for John Gould which established Wolf's name in London, but no less important was the major work he undertook with Daniel Giraud *Elliot on the birds of North America – work which challenged Audubon's dominance of that field. He successfully navigated his career between the rocks of art and the dangerous eddies of science while showing his vulnerability to the mud-banks of Victorian sentimentality. In short, Wolf is a compelling figure for his art, for his science and for the position he created for himself in British society. But first we have to explain how it was that this German bird painter found greatness in a foreign country.

Leaving Germany Behind

The 70 years between the appearance of Frisch's handbook and the completion of Naumann's 12-volume opus were a period of great activity in German ornithology and bird painting. Quite apart from the names already mentioned, there are figures such as Johann Conrad Susemihl and Johann Michael Seligmann, who would reward closer attention, as would those many painters who took on smaller commissions, like Christian-Eberhard Eberlein who worked for academics such as Blasius Merrem. Naumann's rival,

Ambrosius Gabler, deserves better than to feature as an also-ran of history. In fact, all these pictures cry out for re-discovery.

This period of activity faded markedly after 1840. The science of ornithology thrived, but bird painting limped far behind. Germany lacked an entrepreneurial figure such as John Gould to focus this field of work, and, despite the example of Alexander von Humboldt, few German expeditions had set out after him to discover the avifauna of distant continents and thus capture public interest. As Heine remarked wearily the Germans with their philosophy had conquered heaven but the earth was in English hands. To find a public for their work, artists had to look either to Brehm's perennial best-seller, which carried only black and white illustrations or to the hugely popular family magazines which grew up in imitation of Dickens' *Household Words*. But these magazines too carried no colour work.

Historians have long had an explanation for the decline of German intellectual and artistic life at this time. Not only was it a time when the great figures had passed on – Goethe's death in 1832, like that of Hegel in 1830, was felt to mark the end of an era – but the political situation after the meaningless revolutionary year 1830, still more after the failure of the 1848 revolution, had a severely dampening effect on cultural activity. The repression of the Metternich years, mass emigration to the USA, the stultifying effect of that particularism which divided Germany into numerous petty states (396 when Napoleon invaded in 1806, still 39 states after the Congress of Vienna in 1815), the sense of being stranded in a backwater of history while elsewhere dynamic change was sweeping Europe – in short, there was a widespread despair at Germany's situation. Goethe expressed this sense of hopelessness in his famous lines, 'America, you have it better than our continent, the geriatric.'

There's no doubting these general features of Germany at the time but it is far from clear that they prevailed in the life sciences and ornithology. Indeed, these years saw a great increase in the prestige of institutional science. In chemistry and physics but also in biology, German research could look to real achievements. The universities enjoyed prestige and respect and – once the censorship of the years up to 1848 had relaxed a little – people could work with greater openness than in the strongly centralised institutions of England, dominated as they were by the Anglican church. Whatever reasons there might be for a decline in German bird painting, it was not necessarily associated with any decline in science.

One theory is that bird painting suffered from the gradual decline of the branch of ornithology which had been particularly favourable to the painters, namely: taxonomy. 'Splitting' as scientific practice was good for a painter's business: the more species, the more pictures were needed. It's also been suggested that field-work – from which major impulses for bird painting came (and continue to come) – was gradually being side-lined by laboratory science. For comparative anatomy and physiology, large scale pictures had little importance. A significant event was the founding in 1848 of a learned journal for 'scientific zoology', subsequently a leader in the field. Its declared policy was to reject all contributions coming from what was dismissively called 'descriptive zoology'. This could have been the death-blow to field work and the bird painting which came from it but the work of Lynn K. Nyhart has uncovered counter evidence to the effect that university departments of biology, especially in Prussia continued to encourage field work. Indeed, it was this initiative which led to the early establishment of ecology as a scientific discipline in Germany. Nyhart also shows the central role of public museums in encouraging the spread of the biological sciences. These were the years when dioramas came to the fore – museum exhibits which included not only a few individual mounted birds but created entire biotopes. At least in Prussia biology was understood (before it was banned) as 'the study of the organism in relation to its physical environment'. The effect of this on museums opened a new market for bird painters to survive financially. There was no essential reason why bird painters should not have prospered.

Perhaps the negativity affecting other branches of cultural life in Germany was less important for the painters. Even though numerous distinguished members of the active generation of biological scientists were driven into exile – Lorenz Oken, Carl Vogt and Georg Büchner to Switzerland and Louis Agassiz to the USA are a few examples – we shouldn't too readily use cultural pessimism to explain the decline of the attendant art.

Nevertheless, the lives of Germany's two leading bird painters in the nineteenth century represent a clear pattern. While, as we saw in the previous chapter, Johann Friedrich Naumann remained stuck in the provinces, living out in Ziebigk the line from the *Psalms* much quoted in Germany at the time: 'so shalt thou dwell in the land and verily thou shalt be fed', Josef Wolf moved out into the wider world. What Naumann saw as a 'patriotic science' had a quite different meaning for Wolf, for whom it was the key to the world outside

Germany, opening the door to a highly profitable, multinational industry. We have no reason to explain Wolf's emigration to Britain in political terms, yet the contrast between him and Naumann is symptomatic of deeper divisions in the country of their birth.

In the nineteenth century, England was the centre of ornithology, most obviously in terms of money and prestige. Elaborate ornithological expeditions and luxurious bird books found funding only (or at least primarily) in England. The patronage of Lords Derby and Rothschild was coveted across Europe. Searching for commissions, bird painters came to London from Scotland (*MacGillivray) and Holland (*Smit and *Keulemans) and from the USA. In 1848, these talented figures were followed by the young German, Josef Wolf, in whom many see the greatest of them all. One must question whether his reputation would have been so great if had he stayed in the country of his birth. Like Eleazar Albin, unknown in the country of his birth, who emigrated to England a century before and found a place in history as the author of one of the first modern, coloured bird books, Wolf made his career there.

Wolf's Success

Wolf's biographical details are well-known. He was born in 1820 near the River Moselle, where he spent a chaotic childhood, hanging around in the countryside and hunting. He completed an apprenticeship in the renowned lithographical institute in Koblenz (where the Moselle flows into the Rhine). Here his skill and creativity as a draughtsman was recognised and encouraged. The first zoological commissions came from Eduard *Rüppel and Johann *Kaup (a follower of Oken), who worked in the museum in Darmstadt. Kaup was responsible for passing on to Wolf contacts which would have great importance for his future. The first of these was Hermann Schlegel in Leiden, for whose *Traité de Fauconnerie* (1844) Wolf prepared pictures. Schlegel was to be an important mentor and Wolf's great admirer. The second was John Gould, the supreme organiser in British ornithology.

Gould, who did himself not paint, was always on the lookout for painters who could help him realise his ambitious plans. His biographer remarks of Gould's own talents: 'Gould did not himself carry out a single one of his 2990 pictures'. For that reason Gould was more than happy to place Wolf among the illustrators working on G.R. Gray's important *Genera of Birds*. It meant that, shortly after coming to London, Wolf had really arrived. He was soon so well known that David Livingstone commissioned work from him. Wolf's pictures of exotic African animals found a huge public.

It was entirely in the scientific spirit of the times that Wolf described his work for Gray's handbook in the following terms. He wrote in 1849:

> My constant aim is to reproduce as true to life as possible those characteristic variations in the physique of a bird in which ornithologists identify the features which distinguish it from other species.

These words make clear the prioritising of identification as the purpose of the science to which he devoted himself. Wolf thereby initially took over from Schlegel the separation of art and science in bird painting but he drew radically different consequences.

It was not long before Wolf's success made Schlegel's differentiation irrelevant and as he worked with Gould his scientific achievements turned into commercial success. In other respects too, Wolf quickly grew out of Schlegel's influence and adapted to English circumstances while keeping his distance from Gould. He did not care for Gould's rough personal manners, nor did he share Gould's taste for excessively bright colourings in pictures. These were products of Gould's struggle for social acceptance and of his commercial sense and both drove Gould unashamedly towards the market. It was largely for that reason that he became interested in exotic avifauna and in 1861 produced a publishing sensation of the first order: a humming bird book dripping with gold leaf. Early on in their relationship Wolf realised that Gould needed him more than vice versa. Gould had ensured Wolf's entry to the scene but Wolf knew that he had the talent to drive Gould's business and to raise its artistic standards, a recognition which gave him greater freedom in dealing with his one-time patron. Yet it would be wrong to imagine that Wolf himself ignored the market. On the contrary, he showed himself remarkably adaptable in opening up still more profitable sides to it. His contacts to Lord Derby, for instance, had brought him both significant commissions and the opportunity to make lengthy visits to the lord's menagerie, where some of his most detailed portraits were prepared. At the same time, Wolf found that these contacts gave him commissions not only from African safaris but from the grouse moors.

Hunting had not greatly changed since the days of Snyder and Oudry, even if its function as the representation of feudal hierarchy had (perhaps) diminished in importance and the provision of meat had become still more disingenuous as its rationale. In Wolf's Day hunting was an activity for the social and financial elite on the moors of Northern England and Scotland. Oscar Wilde's famous summary of 'the unspeakable in pursuit of the inedible' referred to hunted foxes rather than to pheasants and grouse, but Wolf might have agreed with the 'unspeakable' since neither form of patronage – big game or 'game birds' – did anything to shift his personal loathing of the mindless slaughter of noble animals and his scorn for aristocratic shot-guns slaughtering pheasants. It is noticeable that his portraits of North American pheasants are among his finest work. His clients appreciated his pictures but they appear to have ignored his opinions.

Wolf's success allowed him to restrict the subjects of his paintings to a few species. He had no ambition to produce a comprehensive work and certainly no desire to compile any field-guide. Completeness was low on his list of priorities – a sign of the increasing specialisation of the market and of Wolf's own tastes. Working on individual commissions, he could prioritise the species he preferred. No one knows how he might have painted a starling or a bittern.

Wolf had another talent which contributed greatly to his popularity in Britain, that of portraying the character and mood of birds. We saw in Oudry's work that Wolf was far from the first to have that ambition. Unlike Oudry, however, Wolf was careful to add small but carefully executed landscapes to the portraits. This had the function among other things of showing the bird's relationship to its surroundings and giving a sense of its general behaviour. Critics see here the influence of the landscape painter Carl Ludwig *Seeger, from whom Wolf received instruction in the Grand Duke's drawing school in Darmstadt. But, as the comparison with the work of Bruno Liljefors (to which we return) will stress, we need to examine the type of nature which Wolf's landscapes portray.

The importance of these backgrounds to Wolf is shown partly in the fact that he painted them himself, no minion was given that job. Wolf left behind him almost for good Naumann's portraits without background. We also note that Wolf not only showed his birds as active but concentrated attention on their eyes. This was perhaps a product of his many visits to Lord Derby's menagerie but the activity of his birds came partly from Wolf's experience of the hunt, for the hunt took place in the home territory of many birds of prey and Wolf showed these birds as hunters in relation to their habitat. This was a big advance on Naumann, who had far less chance to observe his subjects in their natural environment. By the Elbe only the smaller birds of prey were observed and many of his subjects reached Ziebigk only as skins. When Schlegel criticised Naumann's pictures, he was wrong to denigrate his understanding of art or nature. What Naumann lacked was a breadth of experience, something which a peasant in the German provinces could hardly hope to acquire. Wolf's work knew no such limitations.

Wolf and Darwin

It was, among other things, Wolf's ability to characterise animals which drew Darwin's attention to him. In Darwin's study *The Expression of Emotions in Man and Animals*, which was first published in 1872, some 14 years after *The Origin of Species*, Darwin was concerned to establish similarities in facial expression and physical reaction between man and animals as they experience particular emotions. His book therefore contained numerous illustrations, including – this was an innovation for a scientific work – photographs. It was no surprise that Wolf was commissioned for this work, not just because of his ability to express character in animals but for his scientific objectivity. Abraham Dee Bartlett, director of the London Zoo, had recommended Wolf directly to Darwin – Wolf was by that time official painter to the Zoological Society – assuring Darwin that Wolf had 'an eye like photographic paper' (times were changing as the metaphor shows). With this commission, Wolf appears to have orientated his painting still more strongly towards science.

It is anything but easy when considering the bird painters of a previous age to be clear about how close they were to a genuine scientific attitude. In particular, there's a danger in accepting as 'scientific' only those elements which are confirmed by the science of today. The majority of painters were laymen in science, incapable of assessing the scientific value of the theories within which they were called on to work but nevertheless full of faith in science. We can see this combination in the famous novelist George Eliot, an exceptionally gifted woman intellectually and widely read. Eliot worked with at least two major scientific ideas. She used contemporary social scientists to help her to understand her own society and – for instance in *The Mill on the Floss* and to some extent in *Middlemarch* – relied on the theories of the German social scientist and anthropologist Wilhelm Heinrich *Riehl. Modern social scientists would be far less likely to consult Riehl. A few years later Eliot used the observations which Darwin had made in *The Expression of Emotions*, a work which she had read carefully while writing *Daniel Deronda*. Eliot explicitly let objective science guide her imagination and this trust in science is far more significant in assessing her work than the question as to whether or not the theories she followed are acceptable today.

This general problem is particularly acute in the reception of Darwin's ideas. It would, for instance, be unreasonable to regard as 'unscientific' those bird painters who remained unconvinced by Darwin's theories. Darwin's opponents – Louis Agassiz, for instance, a follower of Lamarck, or the great medical scientist Rudolf Virchow – were hardly less qualified than Darwin and certainly enjoyed far greater academic prestige as scientists. We might also remember that Darwinism went through a period of decline towards the end of the century. Even in Germany, where Darwin's work was more positively received than in Britain, it would be quite wrong to imagine that only his theories enjoyed respectability or that any artist unmoved by these theories was therefore working unscientifically.

Bird artists had particular reason to be hesitant about Darwin's ideas. As previous chapters have shown, these painters' enthusiasm was concentrated (well, beyond the jungles of taxonomy) on the careful presentation of an individual species. What Swainson expressed in the sentimental language of a creationist – that the individual species had come into the present direct and unchanged from the hand of God – was not that different, for instance, from Lamarck's theory of orthogenesis. In contrast to Darwin, who had postulated an external, 'natural' selection, with the species subject to constant change, Lamarck explained evolution as proceeding from an innate disposition of the species. This meant that each species kept its unique status. In this respect, the bird painters found a more positive message in the science on offer from Lamarck. Furthermore, Darwin's understanding of variation was dismissed by many scientists as a form of hybridisation and therefore as a further attack on the autonomy of the species. Despite these wayward theories, however, we have no reason to regard as 'unscientific' those bird painters who followed them or to give Wolf particular credit for being right just because he went with Darwin.

The situation is complicated by Darwin's ability to misunderstand the purpose illustrations would have in his work. While Wolf's involvement in Darwin's project was a sign of a certain interest in science, it is disturbing to note that Darwin himself had initially negotiated with the publishers of *Animal Lives* to use their illustrations for his own book. He apparently believed that Brehm's anti-clerical opinions might persuade him to pass the pictures over for nothing. In reality, Brehm was interested in Darwin largely out of market speculation, while his illustrators themselves had no interest in science and wanted to appeal to a broad public through the family magazines. Their pictures were aimed at contemporary tastes for anthropomorphic animal pictures. Darwin failed to notice their unsuitability for a scientific project. Was he more accurate in his assessment of Wolf?

This was not Darwin's first problem with his illustrators. He had, for instance, previously entirely failed to persuade artists to present sexual selection as he wished. (We might recall that a discussion of sexual selection by the female bird formed the major part of *The Descent of Man* and *Selection in Relation to Sex*. Birds remained an essential part of his argument long after the famous finches.) Of course, it is not easy to portray the *process* of selection in a static picture but for most illustrators the problem was ideological rather than technical, in that Darwin's theory flatly contradicted Victorian understandings of gender roles. The painters continued to portray the male birds as energetic and full of initiative, and, while this made for attractive pictures, it hardly supported Darwin's theories. A popular science book of the 1830s had described the male bird as 'pascha of a harem' and this cliché gave illustrators a more sympathetic model than Darwin's science.

Similar problems had occurred with the painters' approach to Darwin's idea of the struggle for existence. They were reluctant to paint blood and death even as part of their portraits of birds of prey, and, as we saw, invariably refused to portray the shrike together with its larder. Butchery took place a long way off the page. If pressed, illustrators portrayed a struggle for life between species, never within one species. It was not just Frisch who embraced the gory detail: most German illustrators were less inhibited than the English – witness the Gabler picture of a shrike (**Plate 20**). In this Darwin was on surer ground with Wolf, whose pictures of raptors did not hesitate to show the injuries which their prey had suffered.

The space for misunderstanding did not get less with Darwin's *The Expression of Emotions*. It brought him into a potentially highly ambiguous relationship to the painters. His thesis – that human gestures and facial expressions were a product of evolutionary processes and could be traced to the animals situated on the human evolutionary chain – had considerable affinities with a basic position of much animal and bird painting anyway, namely, anthropomorphism: making animals appear human. Of course, that was the converse of Darwin's argument – he had argued for the animal in humans, not the human element in animals. It could easily be argued that, once again, the painters had misunderstood him. If so, however, the misunderstanding was a more general one, for it affected the entire basis of ethology (a discipline Darwin had not founded, but to which his latest work was close). In subsequent generations, scientists critiqued the whole subject as itself being little more than anthropocentric since it used human behaviour as the interpretative model for animals. Before the century was out, Darwin's methods had been dismissed – notably by the biologist Jakob von *Uexküll – as subjective and 'anecdotal'. Only later did ethology shift its focus from questionable assumptions about individual animals to social and group behaviour and thus move away from some of its early ambiguities. The painters, however, were more than happy to continue with Darwin's line. In Wolf's case, however, the consequences were problematic.

Plate 21: Wolf: *Peregrine Falcons*

Wolf supplied Darwin with four pictures (though in some editions only the laughing ape was used). This type of picture further increased Wolf's popularity, but a longer view of his achievement might question whether it was partly responsible for his compromises with Victorian taste. As part of his arrival in the English scene, Wolf had received the questionable accolade of being praised by Landseer. Landseer, whose picture of Queen Victoria's lap-dog had made him the darling of popular taste (providing the biscuit-tins and tea-caddies of many generations with iconic images), could hardly have been further from Wolf's stated ambitions as a scientist. Even German historians – writing in a country knee-deep in similarly atrocious art – regard Landseer as being 'on the edge of kitsch' while Schlegel, doling out marks to European painters, described

Landseer's work as being 'beneath any recognised standards'. It was dangerous to be praised by him.

As we suggested earlier, Wolf's portraits of birds of prey tended to emphasize the individual, personal quality of his subject. Unfortunately, other works took this trend deep into the worst kind of Victorian kitsch. It would be flattering to ascribe this to his interest in ethology: sentimentality remains a more plausible explanation. It was one thing for Wolf to provide illustrations for an edition of Aesop's *Fables* – after all, the point of fables is that animals behave like humans: Oudry had done the same thing when illustrating La Fontaine – but Wolf did an awful lot of animal pictures with no larger ideas than 'human interest' or, worse still, as illustrations of moral truisms. All this made Wolf a favourite of the Pre-Raphaelites – no great guarantee of anyone's scientific style. When we see his picture of a squirrel peering into a collared dove's nest (title: *Nosey Neighbours*), their choice seems natural. Wolf's situation showed how in the Victorian period bird painting could relate closely both to scientific and anti-scientific trends and how easily high art could accommodate itself to a less demanding public. In fairness to Wolf, however, we might bear in mind a damning comment in the *Saturday Review* in 1895 to the effect that Wolf's painting 'was not wooden enough to please the naturalists' while 'many artists found them too scientific. Schlegel rides again.

Wolf's Scientific Pictures

Wolf authored no individual books. In book form one can find various selections from his work, especially collections of the raptor pictures for which he is celebrated. We take one of these serious portraits for close consideration: the picture of peregrine falcons in Scotland, a picture which appeared in the first volume of Gould's *Birds of Britain* in 1862 (**Plate 21**).

This is an active picture with a certain narrative quality. No greater contrast could be imagined to the static pictures of the early nineteenth century. It shows the peregrine in three characteristic poses, the most prominent of which suggests that Wolf probably studied this specimen at close quarters in a menagerie. Great attention has been given to the markings and to the eyes and the rounded breast makes clear that, whatever else, the painting had no connection with any skins delivered to the artist – as would have been the case for so many other artists. This picture depends on hyper-observation but of course it is a construct, for it's unlikely to find the three peregrines here in such close and harmonious proximity, particularly since there is no indication of any relationship between them. It is the processes of identification rather than nature which has brought them together.

The background is not without interest. In Chapter Ten we suggest a comparison with the environment in which Bruno Liljefors (a considerable admirer of Wolf) portrays his eagle-owl. The comparison emphasizes what is evident in Wolf's picture, namely that Wolf has portrayed the landscape with human eyes and has not attempted to see the landscape as it might be seen by the bird. It is simply a Scottish scene with which readers of country magazines or the hunting fraternity would be immediately familiar. It is a different kind of anthropocentrism to portraying the birds as possessing human features, but it is not that different. Liljefors tries to show a landscape in which humans have no say, it is the bird's habitat and it doesn't even look human. Wolf's is a human world with birds in.

For the critic Heidrun Ludwig, the role of landscape in Wolf's work – a function of his training in Darmstadt – was to shift his work from sober, prosaic ornithological illustration into aesthetically significant bird painting. Her remark presumes a principled opposition between science and art. But the double aspect of landscape – as a humanly appreciated 'nature' and as habitat – seems more significant than a crude separation of this kind. The assumption that the addition of 'nature' was essential for a bird illustration to qualify as 'art' seems little more than a hangover from classical aesthetics (no one seemed to have a problem with Dürer's *Little Owl* in this respect) – and that has to be a result of the fact that 'nature' came into science so much later. We recall that nature, got painted as an interest in identification increased, the bittern's reed-bed or the fish-hawk's seascape had a kind of token quality, which Wolf's pictures finally dispense with. It would take the more focused work of biologists to create an atmosphere in which all animals' special relationship to their habitat might be explored: the idea of the biotope required a different form of representation than nineteenth-century landscapes. The mastery which Wolf displays in this picture is not thrown into question by these considerations. They do, however, set a historical framework for his oeuvre. Not just the Landseer portraits belong to the limitations of their age.

Wolf's Legacy

A suitable way to pay tribute to Wolf is to conclude this chapter with a brief account of three painters who followed the tradition he established, each of them – as it happens – born in 1860.

His more immediate successors were two painters from Wolf's adopted country, who knew him personally. The Scot Archibald Turnbull (1860–1935) was the son of a highly regarded portrait painter. His watercolours are close to Wolf's work, some of them are indicated by the artist himself as being 'after J. Wolf'. Many of his pictures share Wolf's themes: the open country of the Scottish Highlands and the region's birds of prey. But Turnbull was also a talented painter of small birds. He was a pioneer of the protection of birds, encouraged the spread of popular ornithology and worked for choice on affordable (rather than luxury) field guides. At the same time, together with Wolf's second successor, the painter George Edward Lodge (1860–1954) and one other colleague he worked on the coloured plates for the three volumes of Lord Lifford's *Coloured Figures of the birds of the British Islands* (1885–1897), a work which Claus Nissen describes as 'perhaps the most beautiful and most perfect iconograph in ornithology'. Way beyond Gould's circle, therefore, Wolf inspired those working on the finest works of scientific ornithology.

The most innovative of Wolf's successors was a painter who had no personal contact with him and knew his work only from a work Wolf had illustrated on Swedish birds. Bruno Liljefors (1860–1939) – the principal subject of Chapter Ten – shared with Wolf and Lodge a closeness to the hunting scene, even if in the Sweden of his day this was a less aristocratic pursuit than in Britain. Where he moved onwards from Wolf was not simply in his readiness to use photography in his production process, but – as we suggested above – in his treatment of habitat. He takes the integration of the bird into its habitat so seriously that – to pick just one example of many: the portrait of a snipe – the bird is hardly visible in the high grasses. It's an unusual portrait where you have to look very hard to find the subject (Wolf himself once used this technique himself in a portrait of a woodcock). Innovative too is Liljefors' fascination with groups of birds – another feature which greatly reduces the anthropocentric element of his work.

Liljefors' interest in birds' camouflage and in their group behaviour lends his pictures tension and dynamism, overlapping perhaps sometimes with some of the excitement of the hunter. He is a worthy successor to Buffon's illustrator whom we saw observing nature by using the subterfuges of the hunter. Liljefors' well-known picture of a flock of eider-duck taking off from the waves could not be further in spirit from the bird on the branch style of portraiture. The movement of the wings means that detail is reduced, but, as in some of Audubon's paintings, the rhythm of flight – of course a feature distinctive to individual species – is communicated. Chapter Ten looks in detail at the art history trends as well as the moves in science which helped Liljefors in these directions. Here I wish to stress Wolf's importance to Liljefors' development and to identify their relationship as a continuation of the internationalism which has always characterised ornithology.

Before turning to Liljefors and one more painter who shows the influence of modernism on bird painting, we need to take in a final aspect of the representation of nature, a movement whose shadow hung round the topic from the mid-nineteenth century on. Painters and critics alike at times cherished the hope for its revolutionary effect on their work. The example with which the next chapter is concerned is a little later, dating from approximately 1906. But photography cast a long shadow not simply into the present when bird painters might fear for the future of their profession, but backwards into the past.

Chapter Nine
Watch the Dickie Bird: The Bird Picture in the Age of Photography

Early on in the book I suggested calling the pictures we discuss 'portraits'. Some might be better portraits than others, some might be overdramatic, others stiff and wooden, but we have not presented a picture for which the term portrait was inappropriate. And this remains true for this chapter, which discusses the triumphant progress of photography through the second half of the nineteenth century and its impact on bird painting. The title of this chapter comes from a phrase current a century or so ago and now perhaps less familiar. It told the person posing for a photograph to look into the camera by referring to a small mechanical bird which emerged from some cameras just before the shutter was released. In this chapter, however, it is the birds who look into the camera.

The first market which photography wished to conquer was the portrait with the explicit aim of making traditional portraiture redundant. The majority of the early daguerreotypes were in fact portraits and – as an article in a London photo journal of 1861 makes clear – that was no coincidence. We read:

> [photographic portraiture] has in this sense swept away many of the illiberal distinctions of rank and wealth, so that the poor man who possesses but a few shillings can command as perfect a life-like portrait of his wife or child as Sir Thomas Lawrence painted for the most distinguished sovereigns of Europe.

The photo as a democratic portrait. Inevitably, however, the principal subjects of the early photographs came from the old elites. Upper bourgeoisie and aristocracy put their properties before the camera, then their marriageable daughters, then their hobbies. Among their 'possessions', their colonial subjects formed a significant part with the so-called 'ethnographic' portraits slotted in next to the dead lions and other great spectacles – virgin nature, lofty mountains and waterfalls.

No better slogan for the camera's role in depicting nature could have been found than the remark of the English country parson, Fox Talbot, to the effect that the camera was 'the Pencil of Nature' (1844). That set the scene for the repeated claim that in photography nature presented itself. If landscape painters tried to copy nature, the promise was that in photography nature copied itself – an unbeatable offer.

Photography offered much to the sciences too. An early advertising campaign for Kodak emphasized the ability of their products to deliver a permanent and objective archive of the world. Social researchers and collectors leapt at the chance. The criminologist Cesare Lombroso built up his archive of criminal types with photographs and, as we noted, Darwin used many photographs in *The Expression of Emotions*. It was the fact that photos could be endlessly repeated without their content varying (like the best scientific experiments) – this was the 'mechanical reproducibility' of which Walter Benjamin famously wrote – which made them the sister of the exact sciences. It was to be expected that photography would have a major effect on landscape painting and on bird painting in particular.

This chapter focuses on what appears to have been the first handbook made up exclusively of animal photographs published in Germany, Hermann Meerwarth's *Life pictures from the Animal World* (1906). Of its six volumes, three are devoted to birds. Meerwarth had previously published a shorter work offering a guide to nature photography 'for amateurs and friends of nature'. He includes in his collection ideas and samples from the work of the American nature photographer A. Radclyffe Dunmore, but his base was in Germany with the photographic company Voigtländer. Many of the photos in his handbook had won prizes in Voigtländer's annual photo competition – pointing once again to the close links between the advance of science and commercial interests. However, it's Meerwarth's theories on photography which most interest us here.

Modern, Yet Archaic

In his preface Meerwarth praises the new technology and what it has made possible. He writes:

> With every passing day, photography is being recognised and taken up as the only perfectly satisfying method of illustration. Just as in written descriptions of nature we accept only that which strictly adheres to facts that have been properly observed, so the accompanying illustrations must eliminate any kind of imagination in their depiction of nature. What other medium can offer in this regard the same guarantee as an optical system which follows the laws of nature and which prints on the plate only that image which really exists inside the photographic apparatus?

It's a typical eulogy on the camera and the 'absolute' pictures which it produces. The picture is not guided by any human hand or imagination: 'it is made by the things themselves' (Fox). In his own book Dugmore adds the final cliché that 'the camera cannot lie'.

There were not new sentiments. They are reminiscent of the pioneers of the 1840s, such as Edgar Allan Poe, who enthused about the daguerreotype for being 'infinitely (we use the term advisedly) infinitely more accurate in its representation than any painting by human hands'. The desire for exactitude was a dream cherished for longer still by painters. The Scottish bird painter William MacGillivray, had written in 1819 of his desire to 'to copy nature with the scrupulous, yea even servile attention'. In his generation MacGillivray had no option but to stay with his traditional skills. The classical painter Jean *Ingrès too longed for the exactitude which the camera could bring to painting: 'It's something wonderful,' he wrote, 'but we're not allowed to say so.' However, his pupil Charles *Nègre became one of the best known and significant innovators in the new art.

The arrival of photography meant far more than a technological change. Its social consequences – clearly identified in the earlier remarks on the portrait – were hardly less important. It opened up to the broad population a much easier access to the production and reception of images and at a stroke challenged the elite character of the traditional world of pictures. Not only the rich and powerful could commission pictures and enjoy them. Admittedly, it took several decades before the promise was fulfilled. For as long as in Paris a glass photographic plate cost 29 gold francs, photography would remain an activity for the better off, but the democratic credentials of the new medium were recognised from the start. A small indication of this is that Oken's science association immediately welcomed photography as part of the democratisation of science which was its general aim.

Meerwarth's embrace of photography anticipated the spirit of a series of left-of-centre essays on the subject coming out from the end of the 1920s (by then also, of course, devoted to the film) – essays still in discussion today. The French critic André Bazin wrote of the camera's ability to produce images of 'the object freed from time and space'. The camera lens is 'the impassive lens, stripping the object of all the ways of seeing it'. These were exactly Meerwarth's arguments. Yet when we look more closely, we can see that these arguments were already out of date in 1906.

The Camera as Observer

The central point of Meerwarth's argument was the absolute quality of photographs. By 'absolute' he meant both the camera's principled difference to the human eye and its infinite ability to record detail. In his preface he makes clear the superiority of the camera to the most sensitive of brush-work. The photograph has qualities which the human hand can never emulate: 'in the reproduction of soft shapes while achieving the most minute detail, for instance in capturing the hairs on a mammal's skin or the veins in an insect's wing'. Meerwarth is here not only bidding for scientific support, he was tapping into the contemporary interest in close-ups, part of public fascination with the microscope (once called the picture machine). A clear sign of this interest was the book produced by Ernst Häckel, under the title *Art-forms in Nature*. Here Häckel published his own wonderful paintings of the marine microorganisms on which he worked, especially his beloved radiolarians. His view was that nature was the origin of art and that copying of the tiniest structures in nature could be the basis of human art. Häckel, a real double-talent, painted all his pictures by hand. Meerwarth knew where his rivals were. Nevertheless, in three areas his assumption that photography was 'the Pencil of Nature' was already questionable.

a) *Colour and movement* Meerwarth's pictures were in black and white and he has nothing to say about colour. Yet colour photography was being actively discussed at the time and there were procedures for colour printing. Indeed, in a frequently reprinted book from the time, Arthur Hübl was in effect repeating about colour photography all the arguments which Meerwarth had been putting forward for his technology. Hübl writes, 'Every pictorial representation strives to achieve the greatest possible similarity with the object and aims to be able to replace the object.' (This was Gessner's argument and Oudry's justification for painting the menagerie).

Hübl was highly critical of black-and-white photography and in particular sceptical about the claim that it could exactly equate to the object. His argument is interesting in its application to Meerwarth's work: 'Establishing the contours of the object seldom ensures an adequate representation: we use the gradations of shade to create the effect of the object's physical existence and we need colour to bring life and truth into the picture'. In other words, Hübl sees the instrumental function of elements in black-and-white photography, not of representing the object but of communicating through convention with the viewer. What the plate contains is not reality itself but a construct of conventions and compromises with the intention – held in common with other media – of creating an illusion of reality.

It's particularly striking that Meerwarth has nothing to say about film – not that we could reasonably expect him to advertise a competing medium – but film was one of the most eagerly followed developments in photography at the time. It had particular relevance to ornithology in the experiments carried out by Étienne *Marey in France and Ottomar *Anschütz in Austria and the Swede Oscar *Rejlander in London. Through sequences of high-speed images these pioneers had established facts about the mechanics of birds' flight, realistic enough for the aviation pioneer Otto von Lilienthal to use them for his experiments. Meerwarth's problem was not leaving things out – it was that he had exaggerated.

He made the objectivity of natural laws absolute in the camera, yet it is humans who look at photographs. Meerwarth seemed unaware of this problem. August *Strindberg, dramatist and amateur astronomer, was as impressed by the power of the camera as Meerwarth and no less desperate to get away from the anthropomorphism and human self-centeredness which interfered with exact science. He often spoke of the defects of the human eye – 'its perspectivism, its foreshortening, its parallax and so on' – and complained above all at how the human eye wanted to take possession of everything it sees and transform objects in its own likeness. Yet he knew that there was no escape. 'You're trying', he noted to himself, 'to conceive of the world but not in the way your eye takes possession of everything but how can you know what's on the photographic plate when the only way you can examine those plates is through your possessive eye?'

Strindberg was working at a level of abstraction way above Meerwarth – who had little other ambition than to sell his handbook. But Strindberg's doubts help us to see Meerwarth's exaggerations for what they are.

b) *The state of photography around 1906* – a simple look around the state of photography in his own day should have made Meerwarth more careful in his remarks. For developments over the last 30 years of the nineteenth century had shown that photography was consistently moving away from the realism Meerwarth was advocating. We've mentioned that some painters used photography in minor ways. It's truer that photographers spent much time imitating not nature but art. They imitated the Impressionists – whose preference for *plein air* was meant to imitate photography – concentrating on light effects rather than on objects and persons. They copied Pointillism, deliberately introducing out of focus effects and coarse grain into their photos to match its techniques. For all that photography presented itself as the handmaiden of science, it's more accurate to see it involved in a process of aestheticizing reality. More and more it showed itself to be just one medium like any other. In this it was responding to the same consumer pressures as painting – with similar results.

For some nature photographers, what was happening to their medium made them campaigners for the return of realism. The American George E.F. Schulz demanded in his book (translated in 1908) that photography be led back to a 'straightforward documentary style' while Peter Henry Emerson in Britain bemoaned 'the death of naturalist photography' and called for photography to resume its 'documentary' function. Meerwarth is part of this movement, only he fails to reflect on his situation and peddles the idealism of 50 years previously. Walter Benjamin's essay on photography showed that one could see the subject historically and yet postulate a particular connection between realism and the camera. In the 1980s Benjamin's themes were more generally taken up – Pierre Bourdieu, John Berger and Jonathan Crary come to mind – and by then, after 50 more years of propaganda and advertising, the idea that the camera was in itself realistic was merely absurd.

c) *Aesthetics makes a come-back* Meerwarth's exaggerated praise for photography was an understandable response to the continued attacks from traditional artists. The bird painter George Lodge (whom we met in the last chapter as part of Josef Wolf's 'school') while using the camera as a prop to memory had two principal objections to photography. The camera, he argued, had no organic relationship to nature, it came between nature and the observer. He echoed Baudelaire's complaint that 'a mechanical process' had intruded into his relationship to nature.

Lodge's second criticism was that photography turned its subjects to stone. That was, of course, a natural drawback of early photography, a product of the long exposure times required by the photographic materials. Living subjects could not hold a natural pose for that length of time without looking rigid and unnatural. By 1906 Meerwarth's pictures are not 'turned to stone'. His own exposure times were up to 40 seconds but others are much shorter. Incidentally, he was an early practitioner of what became standard procedure in future nature photography, in that he gave full technical details of his pictures. Benjamin was to note the tendency in photography for the interest to shift from the individual 'producer' (i.e., painter or photographer) to the technical process itself and to the camera. Meerwarth is an example of this, but his photos are not made of stone.

It would have been understandable for Meerwarth to have turned his back on these traditionally aesthetic critiques of his work and followed Schulz and Emerson in casting off any aesthetic consideration. He might, like Humboldt or Darwin, have equated beauty with scientific truth and let the aesthetes go on discussing their equivalent of angels on a pin. But in fact, Meerwarth does everything he can to reinstate aesthetics. The photograph's absolute quality may be possible since artists, who are condemned to be 'subjective, creating only according to their individual temperament' and unable to grasp 'the truth of nature', have been cut out, but

> once we had ensured our main aim, a photograph's truth to nature, then – where it is possible – we have endeavoured to achieve a beautiful picture in order to meet as far as possible the requirements of aesthetics.

This is a strange but in fact banal remark. Having banished from his photographs, anything which might stand in the way of his objects' revealing themselves to the camera, suddenly Meerwarth drags in aesthetics as a bonus. It sounds as if he did not want to give up on the prestige which aesthetics brought into photography. It's striking to compare Meerwarth with Frisch, who in the mid-eighteenth-century had not the slightest interest in aesthetics. Frisch painted his pictures according to science and saw no reason to 'justify' them by adding anything. Meerwarth's text is a reminder of how aesthetics had come to dominate nineteenth-century thinking, especially in Germany. Schlegel's essay had been another reminder. At the same time, it's clear that Meerwarth's meaning was trivial. Every 'guide for beginners and advanced' had a section on Beauty. After passing on tips about holding the camera level and not putting one's hand in front of the lens, such guides contained a list of ways to make landscape 'interesting' or to add 'mood' – a crucial step (remarked a guide which had sold 200,000 copies by 1912) if the landscape 'is going to be something more than a dry aping of nature'. Suddenly realism is being seen as a 'dry aping of nature'. No, the camera is more than a box in which only the laws of nature hold sway: it is yet another form of the possessive eye.

When colour photography was finally working effectively, the anxiety for the bird painters started up again. In her definitive bibliography of bird painting (1937) Jean Anker wrote:

> 'The perfection of photographic pictures and their increasing employment as a basis for the pictorial material in ornithological works has in some degree thrown hand-made pictures into the shade.' It takes little imagination to envisage with what despondency Jean Anker would have reacted to the place which film and television have assumed in our day in the presentation of nature and the world of birds.

The bird in Front of the Camera

We pass over the many photographs in Meerwarth's handbook which can be regarded as still-life since their subject is the nests of various birds. They are successful in this, if – when compared to the work of Eric Hoskins some 20 years later – technically unsophisticated. It is noticeable that these pictures do not get over the difficulties which Naumann and Buhle mentioned in their handbook of birds' eggs (1818): 'that most nests are at the top of the highest trees, you feel dizzy just looking up' and 'on the most sheer of cliffs'. Given that Meerwarth could not get to such places either, his photos are pleasant, undramatic and informative.

To illustrate this bird portraits, I have selected the 'Little Tern' (**Plate 22**). It seems particularly successful – the bird is both in focus and focusing its own attention on something out of shot. The bird is clearly alive. Technically the picture is acceptable, and we note that the graininess of these prints comes more from the quality of the paper on which they are printed than from any shortcomings in the optics.

Plate 22: Radclyffe Dugmore: *Little Tern*

The sharp focus of the picture has to do with the relatively shallow depth of field. The lack of focus on the surface of the water reminds us of the Impressionists – but that is merely incidental. The many opponents of photography had a standard argument against the medium that its profusion of detail made any aesthetic unity impossible. One critic commented in 1880, 'The painter cleanses his picture from these disruptive, arbitrary details which add nothing to the picture.' Meerwarth's picture refutes this argument. The shallow depth of field does the cleansing effectively enough and gives the photograph a clear hierarchy of importance, but this quality has drawbacks.

Among the details eliminated in the photo are things previously regarded as essential for the accurate identification of the species, notably the bird's size and colouration, and, to go back to a criterion much beloved by eighteenth-century ornithologists, the toes. The first of these is especially significant in the example of the little tern since even experienced bird watchers find it difficult to distinguish between the various tern sub-species. Crucial in such identification is, for instance, that *sterno hirundo* is identifiable only on the basis of a beak coloured red and black. As we examine this photo trying to find clues of this kind, we notice the truth of what Hübl claimed in his critique of black-and-white photography that light and shade have an architectural function, building up the basic shapes of the object, its three-dimensionality and definition. The beak is an example of this. The actual details – the objective representation of the bird – become subservient to what are in fact conventions of the medium. In fact, this photograph is one taken over from Radclyffe Dugmore and Meerwarth cannot give either the technical details of the camera and exposure, or the exact species, or even the continent in which the photograph was taken. All of that would have made the publication of the picture illegitimate for traditional bird painters. The fact that, as photograph, its publication is legitimate in 1906 says something important about a process we have kept in mind throughout this book: the changing nature of identification.

Two factors are relevant here. The scientific usefulness of the bird picture often conflicted, as we saw, with the overall perception of the bird. Dürer's 'Little Owl' was closer to nature than Frisch's. That was connected with the fact that birds were identified only ever by killing them first. The camera created the opportunity of having a record of a bird without killing it first. The dream of Buffon's illustrator had come true: optics rather than the gun.

The second factor concerns the status of pictures, something which had changed radically since the days of Naumann or Frisch. As the access to information was democratised, it brought an ever-greater number of images in its wake and these images became accessible to the broad population, for instance through the family journals of the mid-nineteenth century, in which the density of illustrations increased noticeably and then decisively through photography. Old pastor Brehm, owning just two bird books, is typical of the earlier nineteenth century and a marker of the changes which were to follow.

Meerwarth's readers were surrounded by images. They had learnt how to read pictures and one picture brought others to mind. If we compare one of the monuments of history painting in the nineteenth century – Gros' *Napoleon on the battlefield of Preußisch-Eylau* – with the news photos which came out of the American Civil War, the change is evident. Gros has to include all possible pieces of information about that battle in his picture, casualties, sickness, the units involved, the landscape, the weather. The viewer of his picture deciphered those separate elements and took out of it a sense of completeness and totality. The American pictures were by contrast fragments, momentary snapshots; and the totality they achieved came from their number alone. The public learned to put together an overall picture for themselves out of the uninterrupted flow of fragmentary pictures – it was this skill which meant that Michael Brady could call the camera 'the eye of history'. Meerwarth's pictures are a symptom of a similar process. In future, species identification would come from a plurality of images. No single image would contain the whole picture.

Not every bird painter and illustrator had understood the role of the viewer in creating their pictures, even though, as we saw, Illiger had pointed to the function of the imagination of the viewer in reading a bird picture. It was too early to forget that, however scientific an illustration might be, it depended on conventionality respected by producer and consumer. The viewer does not come to a picture as a *tabula rasa* but with contextual knowledge, with specific ideas and expectations and – as Martin Kemp argued – with a personal hermeneutic method. When Meerwarth was convinced that photographs were 'absolute' and lacked any conventionality, he shared an illusion with many of his fellow professions. That illusion was one of the principal conditions of their trade.

Chapter Ten
On the Edge of Modernism: Liljefors and *Nolde

The pictures we have so far considered possess a certain continuity. From Dürer to Wolf, realism of one kind or another was the link. Although the word 'realism' would have meant little to the early painters, the way in which they went about their business shares an artistic practice with the later realists. The simple imitation of nature did not always guarantee a realistic representation of the object and it was rather the painters' variously articulated scientific approach which produced a realism of detail.

My approach to bird painters – like the concept of the portrait – was based from the start on discussions of seventeenth-century Dutch painting. Svetlana Alpers' account of this painting understands art to be 'a cognitively driven, if not strictly scientific art' – a direction which could be selected by anyone within western art. Those who specialised in bird painting settled on this style. Their work confirms what the French critic Eugène Fromentin observed in Dutch painting, namely that 'a consciously acquired knowledge is behind these pictures, indeed it does our job as critics in interpreting our analysis of the pictures is wiser than we are as analysts and teaches our eye what it should see'. This was the tradition towards which ornithology could orientate itself while it developed as a scientific discipline.

Not every historian has been so ready to equate science and art as Alpers, least of all in Germany with its powerful Romantic tradition. A few years before Nolde painted our last example, Roland Piper caused a stir with his study of *Animals in Art*, in which he differentiated fundamentally between a scientific and an artistic style in animal painting. The animal in art, he wrote, is a different animal to the scientific animal. Scientific animals lack form and individuality and it is only art which shows us 'how many lions are in the one lion of science'.

This book has attempted to work against such a differentiation by stressing the close affinities in the abstraction processes fundamental to both art and science. For the individual scientific illustrator, as for the conventional painter, the individual specimen of a bird species has to be more than an individual – the finished work must be capable of generalisation and typicalisation; both of these achievements depend on the power of abstraction. Each process demands both knowledge and imagination. Piper's theory must be questioned and the best way is through Humboldt's example. For Humboldt it is, ideally speaking, artistic seeing which abstracts, orders and classifies, and this truth survives the cases where science and artistic performance separate. Dürer's herbs or Oudry's birds are hardly less artistic for being accurately observed. Scientific illustration involves more than a mechanical anatomical cut-away. And, despite his shortcomings, is it reasonable that artists' handbooks invariably leave out Ferdinand Helfreich Frisch but include Naumann? We're not talking of these painters' membership of professional associations but of their admission as artists to the slopes of Parnassus. These slopes seem to me longer and more accommodating than some critics have liked to claim.

Liljefors and Nolde: Preliminary Sightings

The artists discussed in this chapter broke decisively with conventional realism, even if some of their works can be mistaken for it. They represent the two most important artistic movements at the turn of the twentieth century – Impressionism and Expressionism. Both movements involved a conscious rejection of the naturalism which still dominated painting and literature, though we should not forget that Émile Zola was an enthusiastic supporter of the French Impressionists. Expressionism emerged later, and particularly in Germany set itself up as the opposite to Impressionism. Yet what they shared was almost more important than their differences, for they rejected fundamental elements of the world-view which had made realism dominant in the nineteenth century, questioning the assumptions of the positivistic world-view on which that realism had been based.

Our first picture is taken from the work of the Swedish painter Bruno Liljefors and was completed in 1895. We encountered Liljefors briefly in connection with Josef Wolf, but his affinities to German art were far wider. Liljefors completed his training in Germany, worked and exhibited there for some time, notably in Munich. In this he was typical of his generation, for in the second half of the nineteenth century, Sweden, like all the Scandinavian countries, took much of its culture from Germany. This was true of realism but still more of modernist culture and thought. The work of August Strindberg is unthinkable without his German experience and admiration for Nietzsche, while German Impressionism is unthinkable without the work of the Dane Jens Peter Jacobsen.

Above all, however, it was Liljefors' interest in science which makes his work so interesting. His importance to the work of the present-day master Lars Jonsson has often been acknowledged. Liljefors' startlingly original picture *Eagle Owl Deep in the Woods*, however, deserves a place in any discussion of bird painting, regardless of the context.

Our second picture is taken from the work of Emil Nolde. Nolde – associated with that Baltic region of Germany which shares so much with its Scandinavian neighbours – came to prominence in the early years of the twentieth century with landscapes and flower pictures of an intensity hardly less than that of Vincent van Gogh. A few years ago an exhibition allowed one to see Nolde's flowers side by side with those of the celebrated German Impressionist, Max Liebermann. Liebermann's pictures came across as charming but restrained. Our example of Nolde's work may have charm – that is a personal judgment – but it lacks restraint. It comes from the months immediately before the outbreak of the First World War when Nolde had gone on an expedition to the Pacific, where, somewhat improbably, Germany had acquired one of its few colonies: the Kaiser Wilhelm Islands. It was there that Nolde painted disconcertingly intense pictures of the island people and their exotic natural world. There was much less science in Nolde's approach to his subject that in Liljefors'; what he did bring to bird painting was part of a revolution in perception which was to break open the closed scientific world of the previous 500 years.

Liljefors and Nolde's work amounts to a radical break with the nineteenth century but neither of them fell back into Romanticism. Of course, their break with realism was neither total nor general, both for the reason that they were capable of extraordinary detailed true-to-life depictions of nature and in the sense that, outside the avant-garde which such painters represented, the business of selling pictures went on largely unchanged and the carousel of fashion continued to spin. Realism came back with a vengeance in Germany and if we were to be writing a complete history of bird painting, we would have many examples of painters who remained within the continuity of the tradition. For our narrative, the break represented by these two painters is the final stopping point.

Bruno Liljefors Goes to Germany to Learn Bird Painting

Like Josef Wolf, Liljefors, born in 1860 near Uppsala, spent an aimless childhood in the countryside. He enrolled in the Stockholm Academy of Art but left without qualifying, together with his friend Andreas Zorn, who was to become Sweden's leading modernist painter. Liljefors set out to make a name for himself as a landscape and nature painter and went to Germany in the early 1880s as an apprentice – the first of his many visits.

At this early stage of his career, as he funded himself through small commissions, Liljefors developed a local reputation for his ability to portray animals as living, integrated with their natural environment. This reflects his background in informal hunting activities – he joins Naumann and Audubon in making the transition from hunting to scientific zoology. That was the title of a leading Swedish scientific journal: *Hunting and Zoology*, so the combination was felt to be natural. Liljefors' biographer Allen Ellenius stresses the importance of science to the young Liljefors, mentioning in particular Darwinism and ecology as his focus points. From Darwin Liljefors took a sense of the struggle for existence and for the mechanism of sexual selection. For a short while he worked closely with the Stockholm Biological Museum, working on dioramas. Even after leaving the museum he never lost touch with the scientific thinking of his day.

Liljefors worked as a nature photographer, enjoying the kind of close relationship with the celebrated Swedish optics company Hasselblad that Meerwarth had had with its German rival Voigtländer. Liljefors took a great interest in the ornithological photos of Marey and Anschütz – this was to be reflected in his many pictures of birds in flight. In 1895, he published a volume of photos from the famous bird observation site in Falsterbo in south western Sweden. Through these years, Liljefors worked actively for the movement to protect the Swedish countryside and its avifauna. Sweden was (and remains) only thinly populated and it's easy to imagine that the country offered ideal habitats for birds. But the threat to birds of prey was acute. It was regarded as a patriotic duty to defend Swedish nature, even in art.

We noted that the second half of the nineteenth century was a time of intense cultural relations between Sweden and Germany. Scandinavian literature and theatre was fashionable in Germany (think Ibsen), while for the Swedes, Germany (especially in the philosopher Friedrich Nietzsche) was the epitome of intellectual and artistic modernity. Germany represented Scandinavia's philosophical window on the world, and, in a more practical sense, its art market was far greater and far more attractive than Sweden. Liljefors had every reason to go to Germany.

At this time Liljefors' friend Andreas Zorn was on friendly terms with the most important of the German Impressionists, Max Liebermann. Zorn's pictures were exhibited in Berlin in 1892, where Liljefors visited him and the two friends regularly associated with Strindberg. While Irish avant-garde writers were drawn to Paris and Trieste, the Swedes' compass was set on Berlin. In fact, however, Liljefors' initial engagement was to the painter Carl Friedrich Deiker in Düsseldorf.

Models and the Dangers of Over-specialisation

Deiker (1836–1892) was (and remains to this day) not a particularly well-known name in German art. He wasn't as good a painter or as scientifically committed as Josef Wolf, whom Liljefors so admired. But in the second half of the century, Deiker had a solid reputation and his name can easily be found in the catalogues and hand books of the time. Two things are significant for his relationship to Liljefors. The first is that Deiker was known to wish to place himself in the great tradition of animal painters. His greatest admiration was reserved for Frans Snyders and he will have instilled a great respect for that tradition in his pupil.

In the second place, Deiker represented an established category in the art-scene of his time, that of a painter of animals and of the hunt. Liljefors was later to express negative views of this narrowing of an artist's work. For him Deiker's work was a clear sign of an increasing division of labour in the profession, which indeed went so far as to mean that various painters advertised themselves as specialists in particular animals. If a wealthy hostess looked to adorn her reception rooms with pictures of horses, she would look in the catalogues and find her specialist – in this case perhaps J.B. *Zwecker. If she felt that a stag or wild boar would better set off her decor, then she would send for Herr Deiker. That's what the division of labour looked like in practice.

There was another area of painting in which the excess of supply was as striking and it is one to which many of the novels of the nineteenth century draw attention. Novelists needed to look no further than Rome, a city to which – for Germans since the days of *Winckelmann and *Tischbein – hundreds of impoverished artists from all the Northern European nations had flocked, where they read Vasari's *Lives of the Artists* and dreamt of greatness while fighting for commissions from well-heeled tourists who flooded the city. The competition did not lead to a division of labour, for all the would-be masters spent their days producing more or less identical imitations of classical painting and sculptures. At home in Germany, however, competition led to over specialisation.

It could be argued that over specialisation had existed in bird painting at least since 1800. If earlier artists specialised at all, then within much broader parameters. Ferdinand Helfreich Frisch had to cover the whole range of species which his father presented him with and although we may get a sense of the birds he might have liked to specialise on, there were no openings for such work other than falconry. Only generalists could meet the market. Bewick and Swainson too (within the confines of his theories) had a broader range but could – like Oudry and Knip, though with similar restrictions – choose the commissions which most appealed. It was above all Gould who accelerated specialisation and it's no coincidence that painters who worked for him – such as Wolf or Lear – are strongly associated with particular species of bird. Like the situation of the seventeenth-century Dutch painters, the prosperity of their society created a range of individual consumer demand broad enough to fragment.

Liljefors' view of this specialisation can be seen clearly in the fact that his subjects do not include Deiker's standard product lines – wild boars and stags. Whatever else he learned from Deiker, it was not from his positive example. Indeed, Liljefors complained that the market tended to destroy the one value which his painting set out to cherish: the unity of nature. He was not alone in rejecting as 'fatal' an 'academic' perspective on nature. He was interested in the whole of nature, not just in those bits of it from which a living could be made. We should note, however, that the criticism of an academic approach to nature did not involve any critique of science – indeed, as we shall see, it was science which brought nature most alive for him.

Swedish artists working in Germany can be grouped according to their artistic ambitions. The first group preached the need for a return to nature, for simplicity and down-to-earth realism. Liljefors, whom from his early days on critics praised for the 'power and simplicity with which he portrays nature' was close to this group – its spirit is best summarised in the work of the Norwegian novelist Knut Hamsun. A second group to which Liljefors was also close was epitomised in the friendship between Zorn and Liebermann and further intensified in the boundless admiration expressed by the two major German language poets of the time, Rainer Maria *Rilke and Hugo von *Hofmannsthal, for the work of the Dane Jens Peter Jacobsen, especially for his novel *Niels Lyhne* (1880). These relationships established Impressionism in Germany. A last group took shape around 1905 in what became known as Expressionism. The Scandinavian dimension of this group is best seen in the paintings of Eduard *Munch and the explosive family dramas of Strindberg.

Eagle Owl Deep in the Woods

It was from the first two of these groups, as well as from his impatience with academic painting that we can trace the impulses which went into Liljefors' remarkable picture *Eagle Owl Deep in the Woods (***Plate 23***)*. This is our last owl in the book and although it would be hard to find a greater contrast than to the *Little Owl* from which we set out, it is still a portrait, unusual only in exploiting a different type of relationship between the subject and background. Even though some authorities list it under a title which omits the species name, it remains a portrait and grabs our attention as such.

Plate 23: Liljefors: *Eagle Owl Deep in the Woods*

Perhaps it will be helpful to begin by collecting impressions, starting with the powerful sense of colour. In that respect the picture meets fully what a theoretician of Impressionism wrote in 1896: 'Nature has stayed what it always was but where it was closed, grey and overcast, it is now blossoming in colour, vibrant, livid even.'

Despite the luminous colours, the owl itself is entirely realistic in portrayal, its feathers painted with great care. The effect of the picture lies in the contrast between the darkness of the woods and the intensity of the blue night sky. Within that wide range all other colours stand out contrapuntally.

Without overplaying the point, we might recall the intensive discussion of colour in nature which Darwin and especially Wallace initiated within the zoology of the time: colour as part of sexual selection, as camouflage and as mimicry. Liljefors knew these debates, so – surprisingly for those who think that his interests went no further than 'blond beasts' – did Nietzsche. When he writes about Darwin, colour is on his mind. It's not essential to refer to art-history to account for the striking colours in the picture, however, much the blue is reminiscent of Caspar David *Friedrich.

Few pictures seem more remote from the artist's studio than *Eagle Owl Deep in the Woods*. It obviously belongs with the pleinairism which the Impressionists favoured and Liljefors made this link clear in his later extended stay in the legendary home of that movement, the village of Grez-sur-Loing in France. Nevertheless, it's clear that the open, fresh air of his picture has nothing to do with a human activity, as was the case with the French Impressionists. No one is breakfasting here or bathing. Indeed, there is no element in his picture to imply that any human being is enjoying, appreciating or even identifying the scene as 'landscape'. The eyes which take in the forest belong to the owl alone. His size emphasizes the centrality of his vision. The elevation he is on is his alone. Compare this with the carefully painted scenes which form the background to Josef Wolf's birds of prey and it's clear that Liljefors is in no way humanising this scene. It's the bird's habitat – it's clearly being observed but only by the owl. It's no coincidence that Liljefors' next important picture, *Eagle Owl in a Snowy Forest* (1896), shows an owl bringing a hunt to a successful conclusion. The bird's eyes are full of patience, calculation and understanding – they are the same eyes as in the earlier picture. It is not a human hunt, it is the bird who observes.

In his classic discussion of the question *Why look at animals?* John Berger examines photographs taken at the zoo. He suggests that these pictures, which often include the notice with information about the species, together with its habitat and distribution, participate in a process of decentring the animal itself. It's as if the animal is being seen through the thick glass of an aquarium. This 'way of seeing' is, in Berger's view, a response to two human conditions. First, to the technical 'clairvoyance' of our cameras and secondly to the level of information which we possess about the animals we are trying to see. Both alienate us from seeing the animals as they are, they are two forms (there are others) of an anthropocentrism which makes objects out of the animals rather than subjects: the animals disappear. Liljefors' picture breaks the aquarium glass and – though it depicts a particular species – avoids the drawbacks of the informational notice. The world he reveals is not that of the human observer. The owl is looking for itself.

In earlier chapters we reported discussion of the difficulty of painting birds' eyes. We took in Schlegel's critique of Naumann for this aspect of his work and the same criticism was made (admittedly by art-historians) of Dürer's *Little Owl*. I know of no similar critique of Liljefors' owl – indeed, it would be a challenge for anyone to make it. The owl's eyes are painted naturalistically, they are neither too large nor too small but they hold our gaze steadily while they take in the woods. As we look at the picture, we feel we are part of the owl's territory – we, not the owl, are intruders. In an enigmatic remark, Liljefors commented that animals should enter a picture 'like a violin in a Beethoven violin sonata', in other words, the landscape is the bird's before its entry. When the philosopher Ernst Mach spoke of an *Empfindungskomplex* (a cluster of sensations) in Impressionism, he meant the sensations of the artist/observer. It's disconcerting in Liljefors' picture to realise that it is the owl whose impressions are being registered.

The picture created a sensation when it was first shown; among other things, there was public indignation that the state had not immediately purchased the picture for the National Museum. It became clear that the search for patriotic symbolism which Dürer had avoided was far from dead. Plenty of critics expressed the conviction that the owl represented the wildness of Nordic nature, others saw the owl as an incarnation of one of the mythical figures which Nietzsche's work had popularised in Sweden. So the owl was said to sit like the 'Great Pan', a nature king on his throne with the sky as its baldachin. It was as if Liljefors' picture was being put in an up-dated emblem book with Nordic paganism as its inspiration. Liljefors was far from happy with this trend and vigorously rejected any links to Nietzsche. Such ideological readings of the owl as 'the daemon of Nordic nature' were unwelcome to him for two reasons. First, they forced Impressionism into symbolic or allegorical messages. For Liljefors and his friends, Impressionism was about closeness to nature and a down-to-earth realism. Secondly, as we see in the following section, such interpretations quite missed the scientific origin of Liljefors' world view.

Liljefors and Science

The relevance of the argument I wish to put forward about Liljefors' picture is not something I can prove – it simply seems to deal with the picture's elusive quality much better than the art-history books. To explain the picture with reference to Liljefors' divorce and the depressive state in which he painted it in a Copenhagen hotel room does not seem to touch any of its features. Liljefors' personal mood seems the least relevant element in the work. But – without forcing the issue – I'd like to sketch a rather different background and to offer a brief account of what science meant to Impressionists in his generation.

This is best approached through a text written in 1910, some 15 years after the picture was completed, by the German doctor and poet Gottfried *Benn. It consists of a discussion of the work of the Danish writer Jens Peter Jacobsen. It reflects on the widespread enthusiasm for Jacobsen's work which followed from his novel *Niels Lynhe* and it endeavours to explain Jacobsen's work not as a literary phenomenon – which is how his reception by Rilke and Hofmannsthal presented it – but to tackle the question: what does Jacobsen's art have to do with his science?

For Jacobsen was not simply active as a novelist, his principal activities were scientific. Not only did he translate Darwin into Danish but throughout his life, Jacobsen had worked as a microbiologist, conducting his own research, which had focused on tiny microscopic organisms, green algae called *Desmediazeen*. Benn's text emphasizes Jacobsen as scientist in a lab coat, bending over his microscope and concentrating on that radiantly illuminated drop of water in which the *Desmediazeen* are to be found which are his subject. It's from this scientific world that Benn traces the origin of the compelling images of Jacobsen's novel, images in which Niels Lyhne lapses 'into a remarkable vegetative state' and in which he experiences a totally passive openness and receptivity to nature – an openness which is so strong that it all but extinguishes his own will and personality and from which reason offers no way back. The passive openness to nature impressions is classed as Impressionism, though it seldom is portrayed in such an extreme form. Benn demonstrates that it comes directly from Jacobsen's Darwinism.

The argument goes like this. Darwin had gone so deeply into the driving forces of all existence – not just the 'natural world' but also of human beings – that it had become impossible for those who followed his work closely to draw a line between themselves as researchers and the objects of their research. Of course, in small ways Darwin did this himself, but we know only of fairly pragmatic situations where his science impinged on his personal life, for instance relating to his children's behaviour. Behind his experience of 'the entangled bank' (in the *Origin*) we can sense a feeling similar to Jacobsen's. The

Darwinists, however, drew consequences from such moments (Darwin was more than used to his followers going further than he did). While public debate focused on what Benn's text regarded as 'childish' questions – for instance, man's descent from the apes – thinkers like Jacobsen understood the demonstration of evolution as evidence of a far more intimate relationship between humans and the whole of nature – of which they were a part. All the things Darwin had observed – the struggle for existence, the hidden strategies which life adopts in sex selection, the evolutionary continuities which make humans allied to all other life-forms – made it impossible for Jacobsen to distinguish between his own life and that of the tiny life forms he sees under his microscope. The sensitive, technically trained scientist felt such affinity with the *Desmediazee*n that there was no way out of his problem, no classification system, no objective descriptions, no belief in human beings' exceptional place in the universe which could restore his sense of his superiority to nature. Subject and object of science are, Benn concludes, 'intimately related, both belong together through both *one* wave flows: both share one body, right down to the chemical composition of their body fluids'. One cannot experience as if one is separate from nature.

That's where science brought Jacobsen. In Berger's metaphor, science no longer builds the thick glass of the aquarium but finally has shattered it. Man at one with the whole of nature.

That's how Benn's text summarised the science of one of the most influential Impressionists of the late nineteenth century. Other essays which Benn wrote at that time convey something of the effect of this change on visual artists, notably on Vincent van Gogh. By then, art fashions have moved on to Expressionism. Yet it's hard not to be struck by the account Benn offers of Darwin's impact – once, that is, the 'childish' polemics and the initial sensation-mongering have passed. The extraordinary openness to impressions from the natural world outside, the loss of science as a dividing wall opened up new dimensions and for the artist in the scientist it raised the vital question, namely how this new experience of the world was to be portrayed. It's hardly possible to use the conventions in words and pictures in which previous generations had expressed and experienced the idea of man as the high point of creation. 'Do you really believe,' Benn's text demanded, 'that you can get anywhere with the words on offer to you from the past: pale, clapped out, tired words?' Science has revealed something which demands a fundamental rethink in art. Where would we expect this to be more keenly felt than in a branch of art whose best practitioners had for 200 years endeavoured to let science into their art and let scientific discovery lead their brush? Liljefors' picture is anything but anthropocentric and has as little to do with personal depression and divorce as *Niels Lyhne*. The picture is anchored in modern science and tries to find a language adequate to it.

If Benn was right in his understanding of Jacobsen, then there is a challenge to the present. After all, the Copernican revolution which Darwin's work represented for Liljefors' time has been followed up by a series of major shifts in scientific paradigm. We need only to consider present day understandings of the environment to see how far changes have gone. It seems reasonable to ask what changes in art may be ushered in by the advances in scientific knowledge of the last century. It's clear that the changes I suggest in the background to Liljefors' picture took place within the artist's understanding of natural life and the loss of the superior position of man in the universe and nothing much was said about the owl itself, except that it should not be seen anthropocentrically. This process has gone much further and one wonders if we haven't reached a point at which scientific understanding has finally become inaccessible to the artist. Anyone who read Jim Al-Khalili's account of recent discoveries about the robin must have had thoughts of that kind. It turns out that the robin on its migration South from Russia, navigates by means of a principle of quantum mechanics hardly understandable to lay people. It seems as if Schrödinger's dream has come to pass, namely that quantum physics would establish itself not just in the debates on the behaviour of elementary particles but would be identified at work in larger processes in 'nature'. Of course, there is no suggesting that the robin understands quantum physics, any more than the chameleon understands the importance of wave lengths in influencing the colours which it changes at will or the kestrel can explain the principles of aerodynamics which enable it to hover but the robin's new status raises questions. Previous explanations of birds' migratory strategies had stayed firmly within lay people's basic understanding. If migrating birds followed the stars, then so did humans. If they used a magnetic North to orientate themselves, then that showed just how natural human beings were in doing the same with their compasses. And even the cranes – who folklore suggested found their way on migration routes by dropping the stone they were believed to carry with them at all times and listening to whether it fell on land or on water – were operating within some kind of appreciable human cunning. But when the robin uses techniques which the majority of humans neither now nor for some long time to come will remotely understand – then things are different. Will that mean that the robin needs to be portrayed differently? Can he feature on Christmas cards as freely as at present, sitting on a spade handle as if he is grateful to the superior species, human beings, for finding him worms and grubs?

This may be speculation but it is not absurd. Frisch and Naumann rejected speculation, yet they launched themselves on the path of scientific objectivity without being sure where it might end. Liljefors stands at the end of 400 years in which bird painting moved closer to science but no one should believe that that process has stopped.

Ernst Nolde and the Attraction of the Exotic

With our concluding account of the work of Ernst Nolde (1864–1956), our discussion moves 20 years forward up to the very last moment of the 'long nineteenth century'. His picture *Tropical Forest* contains no owl but two red parrots and these birds mark the circle which ornithology had travelled, for parrots were among the earliest exotic birds to come to Europe (and to be painted, for instance by Franz van *Mieris), thanks to the import business carried on by the Dutch East India Company. Yet in Nolde's picture, these birds are not presented with the realism of the Dutch painters, nor with the sensitive eyes of the Impressionist, taking in each nuance in the scene 'like photographic paper' – let alone with the careful portraiture of Edward Lear. Nolde's eyes are those of an Expressionist, who is determined to project onto the external world – in this case onto the parrots but also onto the entirety of the forest – his own feelings.

We recall that exoticism had once before posed a threat to mainstream ornithology. It was latent in the conflict between cataloguing native birds and being aware of other species outside Europe and this experience revealed a simple truth that the exotic can easily be reconciled with the normal (after all, what is exotic to one continent is normal to another). Even Frisch reduced the exoticism of the cassowary by breaking his own rules and taking it up into his handbook. By the end of the eighteenth century, birds from outside Europe threatened to swamp the discipline. Buffon feared for the future of his subject, yet the threat passed, largely because of the considerable efforts of the taxonomists to identify and then integrate the unknown species into their lists of the known. Something of the success of these efforts can be seen in the smooth way in which Australasian fauna and avifauna got absorbed by European science – despite their quite extraordinary and 'abnormal' forms.

Precisely these efforts of scientists came under criticism in the second half of the nineteenth century. Not only the scientific expeditions conducted within the framework of colonialism – we think of the travels of the ornithologist François *Levaillant in sub-Saharan Africa – not only the over regimentation of discovery by some of the taxonomists or in Gould – even Linné himself came to be seen as promoting a dry rationalism, an anthropocentric 'mastery' of nature and creating a world without imagination. It could therefore be assumed that the exotic would make a comeback and celebrate the 'return of the repressed'.

Whatever about the influence of literary and philosophical thinking on painters, one artistic movement had been influential in European modernism's approach to nature painting and thus in some ways acted as the precursor of Nolde's picture. I mean the art of China and Japan.

European modernism's discovery of the art of the far east had a long history. It goes back much further than the moment when Captain Glynn sailed into Nagasaki harbour. In the eighteenth century it was fashionable for the wealthy to use tapestries and wallpapers from China, containing colourful and exotic depictions of birds and plants. Their style fed into other decorative arts. The major exhibitions of far eastern art were held in European capitals, starting in Paris (1863) and Vienna (1873), and they awoke a strong interest among the public and a desire to imitate oriental artefacts. In porcelain, such imitations had long been familiar. In music and poetry there were superficial imitations but landscape and bird painters reacted most positively to this wave of exotic art.

There are clear signs of this in Impressionism. Oriental art speeded up painters' abandonment of naturalism and promoted a simplification and stylisation of painting, a reduction of unnecessary detail and encouraged painters in associative, nuanced understatement. In bird painting the so-called 'flower and bird' style of Japan (kachô-ga) drew much attention, for this consisted of juxtapositions, for instance of the purple heron with the hibiscus, the pheasant with prunus blossom. Even the 'patriotic' painters coming from a local tradition found affinity to these pictures for they too wished their pictures to express the inner essence and affinities within nature, nuances which were not to be buried in naturalistic detail. In fact, looking at the humming bird pictures produced for John Gould or of the wonderful American painter Martin Johnson *Heade, it's clear that the synergy of bird and flower was a central feature of many artists' intentions. Audubon too produced many paintings in which this synergy, whether as habitat, affinity or decoration, is the central feature.

It was not just Impressionism which drew on the productive impulses of Chinese and Japanese art, the expressionists also received many and varied impulses from this source. Vincent van Gogh owned and greatly admired copies of the work of the famous Japanese bird painter Hiroshige (1787–1858). Something of this interest can be seen in the painters of the *blauer Reiter* group and in this connection critics have often drawn attention to the 'oriental', passive quality of the animal pictures of Franz Marc.

A further well-known exotic element shaped Expressionism and brought openness to impressions to a new intensity. This was the exoticism of travel, made fashionable by the French painter Paul Gauguin. His extended stay on the island of Tahiti became an iconic statement for his generation of painters. His journey into 'primitivism' combined the exotic with a biting critique of western civilisation and linked in strongly with the widely held anti-rational views we mentioned above. The South Seas had operated as a vehicle for these sentiments from the time of Herman Melville (and his German admirer, Friedrich Gerstäcker) in the 1850s. It was Gauguin, however, admiringly described by Strindberg as 'a savage who hates our whingeing civilisation', who opened this door to European artists. Not every artist was as strong-minded as Gauguin in breaking their links to Europe. When Nolde set out for the South Seas, his journey involved significant compromises. The last stage of his journey was on board a German gunboat. Nolde did not just remain under the protection of Germany as a colonial power but even in the forests he kept a revolver at his side ready to use. And his wife – similarly equipped – was also at his side. He wanted to visit the primitive and exotic but he had no intention of falling victim to them. It was cheaper to scorn European culture – 'Europe, this piece of snot / in an altar-boy's nose' (Benn) – than to turn one's back on the advantages the continent offered.

Despite these ambiguities, Nolde shared with Gauguin the absolute determination to reform art radically by exposing it to primitive forces. Like other artists in his circle, Nolde wanted to put an end to academic painting and to return art to 'the simple life'. He shared with Liljefors an insistence on 'down-to-earth' attitudes and his vision of an art 'down to earth and basic art, redemptive' – saw affinities in the African and Oceanic artefacts which Nolde had been able to visit in Berlin's ethnographical museum (artefacts now constantly in the news since they are in the process of being given back).

One final element should be mentioned in connection with the exotic. Artists' fascination with primitive cultures and their art would never have become so widespread without the work of Sigmund Freud and other psychologists. Not only did Freud give a huge boost to interest in the unconscious mind, in his *Interpretation of Dreams* he analysed some of the iconic representations of birds in legends, classical mythology and the Bible. He also gives an account of various dreams which he and his patients had experienced and which included birds. One did not have to read Freud to be aware of his ideas, for, like Jacobsen's views a few years earlier, they were much in vogue as Nolde set out on his journey.

In a famous essay *Expressionism in Art*, a young German writer drew a radical distinction between Expressionism and Impressionism. In contrast to France, where 'every revolutionary stands on the shoulders of his predecessors' (he's referring to Zola's championing of the Impressionists), the critic postulates a complete break. Artists of the new movement, he writes, 'did not reproduce the slightest thing which excited them, they did not reproduce naked facts. Feeling unfolded itself boundlessly in front of their eyes. They did not see, they did not photograph, they had visions.'

Nolde's picture is one such vision.

Tropical Forest (1914)

The red parrots represent, we suggest, a comeback for exoticism. However, Nolde had not sat at home studying the exotic, he had headed out for distant lands in order to make personal acquaintance with the exotic. That is not, to say, however, that he travelled without luggage.

As we look at Nolde's picture (**Plate 24**), two things strike the eye immediately. First, we notice that Nolde has taken over the powerful colours which Liljefors had favoured. Nolde's inspiration is likely to have been van Gogh, whose work Nolde would have got to know from a major Berlin exhibition in 1891. Van Gogh used the intensity of colours – for instance in his famous sunflowers (an obvious model for Nolde's own flowers) – to dissolve normal perception, to let the sunflowers speak without the word 'sunflower' coming in between. Beyond names, these colours let the intensity of the painter's emotion flood the subject. Here we have left behind any purpose of identification of the flowers (or birds): the painting is about the artists' self-identification with them.

Plate 24: Nolde: *Tropical Forest*

A second immediate impact of Nolde's picture comes from the exuberant vitality and life of the forest ferns and trees. The huge ferns in the middle ground express that vitality – in combination with the red birds we gain the sense of pulsating energy. The naivety of the painting emphasizes that the life force is not held in check by any external forms.

Nolde's painting has left behind the world of Henri *Rousseau, whose many pictures of tropical jungles had rapidly become his signature theme. Rousseau uses similar plant shapes to indicate the vital energy of his jungle but – apart from his greens – his colours are more restrained and he only occasionally uses the passionate red which Nolde reserves for his birds. Rousseau's brush work is more clearly disciplined than Nolde's, despite his preference for naive shapes. His jungle is disquieting, but it can be read. Nolde's jungle is both simpler and less easily deciphered.

From time to time the Expressionists used traditional, mythical figures to evoke the power of the primitive. These are often taken from classical mythology. It was these works which would have fulfilled critics' earlier expectations of Liljefors' picture for the Great Pan is often found there. Once again we can use Gottfried Benn's work and his beautiful South Sea poem *Palau*.

In Benn's poem, as part of the 'dark forest dreams', panthers (the animal of Dionysus) 'leap silently through the trees'. Rousseau too paints mysterious animals in the long grass of his *Equatorial jungle* (1909), animals owing little allegiance to zoology and everything to mythology, while Benn hears 'the cries of the birds of death, the ticking of the death watch beetles, soon it will be night and the lemurines will emerge'.

In *Palau* – famous first line 'Red is the evening on the island of Palau' – the red is not about passion and blood but about sunset and decline. Benn saw Palau as a symbol of cosmic decline and transience. Nolde sees only vitality. The lemurines are the souls of the dead, the ticking beetles mark the end of civilisations, just like Easter Island, whose huge and desolate stone figures stare out eyeless over the unceasing waves. One day Europe too will consist merely of ruined monuments – a thought echoing through T.S. Eliot's *Wasteland* (1922).

Nolde's picture has an extraordinarily positive effect, all the negative aspects Benn and others fix on such scenes are absent. The more we try to put his picture into the general cultural context of European modernism, the more straightforward and unliterary Nolde's subject appears. The cultural references do not seem to fit. The picture evokes the vitality and energy which can be found and which, if Western civilisation would stop repressing it, promise renewal and hope. This hope is a 'down to earth' hope, as all of Nolde's work and it too fits in with the early Expressionists' turning towards 'a basic and truthful life' as with the optimism which came out of Darwin – the promise of fulfilment in participation in adopting a life-cycle which has been going on for millions of years.

The two parrots represent this free and unproblematic natural life. They're no harbingers of doom, certainly not Odin's ravens – they stand for the joy of life, just as – in one of the earliest and most enduring pieces of bird symbolism – birds had represented the human soul and the dove the spirit of God. In a secular world, where organic life has replaced metaphysics, Nolde's birds carry the same message. This was the fundamental content of bird painting through the ages and in every country. It is why the fascination of bird painting has been maintained across the centuries.

Notes

Preface

Derek Niemann: *Birds in a Cage.* Warburg, Germany, 1941. London: Short, 2012 (reference to Stresemann, p. 112f). Angela Wulf: *The Invention of Nature. The Adventures of Alexander von Humboldt. The Lost Hero of Science.* London: John Murray, 2015. Tim Birkhead: *The Wonderful Mr Willughby. The First True Ornithologist.* London: Bloomsbury 2018.

I list here six general accounts of the literature on bird painting. The first two have the most complete, if not exactly up-to-date bibliography: Jean Anker: *Bird Books and Bird Art: an Outline of the Literary History and Iconography of Descriptive Ornithology.* Copenhagen: Levin & Munksgaard, 1938; Claus Nissen: *Die illustrierten Vogelbücher. Ihre Geschichte und Bibliographie.* Stuttgart: Hiersemann, 1953; Christine E. Jackson: *Bird Etchings. The Illustrators and their Books 1655–1855.* Ithaca: Cornell UP, 1985. The last three are the most profusely illustrated: Christine E. Jackson: *Bird Paintings. The 18th Century.* London: ACC, 1994; Maureen Lambourne: *The Art of Bird Illustration.* London: Collins, 1990; & Jonathan Elphick: *Birds. The Art of Ornithology.* London: Natural History Museum, 2004. A more general study of zoological illustration: Claus Nissen: *Die zoologische Buchillustration. Ihre Bibliographie und Geschichte*. 2 vols. Stuttgart: Hiersemann, 1978. More generally on the theory behind illustration: Wolfgang Kemp: *'...einen wahrhaft bildenden Zeichenunterricht überall einzuführen'. Zeichnen und Zeichenunterricht der Laien 1500–1870. Ein Handbuch.* Frankfurt a.M.: Dumont, 1979; Martin Kemp: *Visualisations: The Nature Book of Art and Science.* Oxford: OUP, 2000; K.B. Roberts, The Contexts of Anatomical Illustration, in: Mimi Cazort, Monique Kornell, & K.B. Roberts (eds.): *The Ingenious Machine of Nature. Four Centuries of Art and Anatomy*. Ottawa: National Gallery of Canada, 1996.

For many years the standard history of ornithology was: Erwin Stresemann: *Die Entwicklung der Ornithologie. Von Aristoteles bis zur Gegenwart.* Berlin: Peters, 1951. Much to be recommended: Paul Lawrence Farber: *The Emergence of Ornithology as a Scientific Discipline 1760–1840.* Dordrecht & Boston: Reidel, 1982. Bringing these works up to date: Tim Birkhead, Jo Wimpenny & Bob Montgomerie: *Ten Thousand Birds. Ornithology since Darwin.* Princeton & Oxford: Princeton UP, 2014. On the place of ornithology in biology: Ernst Mayr: *This is Biology.* Cambridge Ma.: Belknap Press, 1997.

On classification: Harriet Ritvo: *The Platypus and the Mermaid and other Figments of the Classifying Imagination.* Cambridge Ma.: Harvard UP, 1997

Chapter One

On Dürer: Colin Eisler: *Dürer's Animals.* Washington & London: Smithsonian Institute, 1991; Sebastian Killermann: *Albrecht Dürers Pflanzen und Tierzeichnungen und ihre Bedeutung für die Naturgeschichte.* Strassburg: Heitz & Mündel, 1910 (here the lament that Dürer did not paint any eagles); Marion Agathe: *Das Bild des Hundes in Albrecht Dürers Gesamtwerk. Darstellungen und Deutungsversuche.* Bochum: Brockmeyer, 1987; Fritz Koreny: *Albrecht Dürer und die Tier-und Pflanzenstudien der Renaissance.* Munich: Prestel, 1985; Ulrich Jenni, Vorstufen zu Dürers Tier und Pflanzenstudien, in: *Jahrbuch der kunsthistorischen Sammlungen in Wien*, NF xlvii (1986–87), pp. 23–37. (Here the lament over excessive symbolism.)

Rousseau quoted in: Joseph Beach Warren: *The Concept of Nature in 19th Century English poets (1936).* New York: Russell & Russell, 1966, p. 138. Horst Bredekamp, Denkende Hände. Überlegungen zur Bildkunst der Naturwissenschaft, in: *Von der Erfahrung bis zur Erkenntnis. From Perception to Understanding*, ed. Monika Land & J. Mittelstraß. Schering Research Foundation Workshop, supplement 12. Berlin: Springer, 2005, pp. 109–32. Goethe and Schiller are mentioned in this context in any German cultural history.

Alexander von Humboldt: *Views of Nature* (1807). Ed. Stephen T. Jackson & Laura Dassow Walls, translated Mark W. Person. Chicago: University of Chicago Press, 2014. Jacobsen's double gift is described in: Morten Høi Jensen: *A Difficult Death. The Life and Works of Jens Peter Jacobsen*. New Haven & London, Yale UP, 2017, p. 37.

Kemp's remarks on illustration in: Taking it on trust: form and meaning in naturalistic representation, in: *Archives of Natural History* (1990) 17 (2), pp. 127–88.

Leonhard Baldners's text was published, unfortunately without its pictures: *Das Vogel-Fisch-und Thierbuch des Strassburger Fischers Leonhard Baldner aus dem Jahre 1666*, ed. Robert Lauterborn. Ludwigshafen a. Rhein: Hofbuchdruckerei Lauterborn, 1903. The manuscript, including the hand-written commentaries of Willughby and Ray, can be consulted in the British Library (Ms. No. 6485). See Christine E. Jackson: (1985, p. 247f): Appendix A: Continental Illustrated Bird Books published to 1660. Above all: Birkhead (2018).

On the Dutch flower pictures: Svetlana Alpers: *The Art of Describing. Dutch Art in the 17th Century*. Chicago: University of Chicago Press, 1983; Gerhard Langemeyer, Die Nähe und die Ferne, in: G.L. & Hans-Albert Peters (ed.s): *Stilleben in Europa*. Baden-Baden: Staatliche Kunsthalle, 1980. It is possible that one could use terms from these texts such as 'zoological still-life' (some critics describe the flower pictures as 'botanical still-life') or 'scientific naturalism'. My choice of the word portrait wishes to emphasize the intrinsic value of the subject.

The outstanding study of Darwin and pictures is: Julia Voss: *Darwins Bilder. Ansichten der Evolutionstheorie 1837–1874*. Frankfurt a.M.: Fischer, 2007. That does not mean that I agree with her in dating bird 'portraits' only from Wolf and Landseer (p. 285).

On bird pictures in medieval manuscripts see: Brunsdon Yapp: *Birds in Medieval Manuscripts*. London: The British Library, 1981.

Pontius Pilate: Brecht gave this line to Azdak, the central character in the *Caucasian Chalk Circle*.

On the bird-watching emperor Friedrich II **see:** Sebastian Neumeister: 'Da es dir gefällt, o Liebe'. Die Dichtungen der Staufer. Winter: Heidelberg, 2021. Also: 'Von der Kunst mit Vögeln zu jagen.' Das Falkenbuch Friedrichs II. Kulturgeschichte und Ornithologie, ed.s Mamoun Fans & Carsten Ritzau. Mainz: Zabern, 2007. This account of Friedrich's inventory is interesting for the recognition that these ancient bird-watchers didn't get their ornithology wrong: climate change and shifting locations of species should be factored into the interpretation. On Marcus zum Lamm: Ragnar Kinzelbach & Jochen Hölzinger (ed.s): Marcus zum Lamm (1544–1606). Die Vogelbücher aus dem Thesaurus Picturarum. Stuttgart: Ulmer, 2000.

Fuller details of the literature on bird-symbolism are given in the notes to Chapter Two. A seventeenth-century brief taster on the owl: 'Minerva's bird on wisdom ever bent / is mocked by those on lighter thoughts intent'. Quoted in: Christine E. Jackson (1994), p. 63.

On Bewick's bull: Jenny Uglow: *Nature's Engraver. A life of Thomas Bewick.* London: Faber & Faber, 2006. The Simon Schama quotation is on p. 168.

A fine study of the plant books: Olaf Breidbach: Bilder des Wissens. Zur Kulturgeschichte der wissenschaftlichen Wahrnehmung. Munich: Fink, 2005.

Chapter Two

On Rembrandt's self-portraits: H. Perry Chapman: *Rembrandt's Self-Portraits. A Study in Seventeenth-Century Identity*. Princeton, N.J.: Princeton UP, 1990. On Dutch painting: Alpers (2007); on the general background: Simon Schama: *The Embarrassment of Riches. An Interpretation of Dutch Culture in the Golden Age* (1987). London: Harper Press, 2004.

It seems that the bittern got confused not only with the owl, but rather implausibly perhaps with the porcupine too. See Isaiah xiv: 23 & xxxiv:11.

Emblem books originate with Andrea Alciati (1492–1550) and his *Emblematum libellus* (1531). Two titles on the emblems in Rembrandt's age: the first a standard work and a more recent book focusing on bird symbolism: Arthur Henkel & Albrecht Schöne: *Emblemata. Handbuch zur Sinnbildkunst des xvi. und xvii. Jahrhunderts*. Stuttgart: Metzler, 1967; José Julio Garcia Arranz: *Symbola et emblemata avium. Las aves en los libros de emblemas y empresas de los siglos xvi y xvii*. Coruna: Sielae, 2010. More general: William B. Ashworth Jr., Emblematic natural history of the Renaissance, in: N. Jardine, J.A. Second, & E.C. Spavy (ed.s): *Cultures of Natural History*. Cambridge: CUP, 1996, pp. 17–37. Also Yapp (1981), who explores further names for the bittern. A small example of Bible ornithology: The Rev. John George Wood: *Birds of the Bible*. London: Longmans, Green & Co, 1887.

For modernism's use of ancient flower symbolism see: Gottfried Benn's poems on the aster and the asphodel: *Gedichte in der Fassung der Erstdrucke.* Frankfurt a.M.: Fischer, 1982, pp. 21, 268. On Rilke's flowers see *Neue Gedichte* (1907).

Eating taboos: Leviticus xi, 13, 24. On dietary matters note also that 'meat-days' referred to four-legged animals, not birds. On Baldner see notes to Chapter One.

The canonisation of the pelican: Lucienne Portier: *Le pélican: historie d'un symbole.* Paris: Éditions du cerf, 1984.

The German scholar is Cornelius Agrippa. On modern aspects of the Melancholia-tradition, together with a useful summary: Mary Cosgrove: *Born under Auschwitz. Melancholy Traditions in Post-war German Literature.* Rochester, N.Y.: Camden House, 2014. Chapman (1990) discusses melancholy in Rembrandt.

On Darwin's eating club: Ritvo (1997), p. 207.

On the gamepiece, with particular focus on Rembrandt's picture: Scott A. Sullivan: *The Dutch Gamepice.* Woodbridge: Boydell, 1984; more general: Claus Grimm, Das Jagdstilleben, in: Langemeyer & Peters, 1980, pp. 253–63; Sarah R. Cohen, Life and death in the northern European gamepiece, in: Karl A.E. Enekel & Paul J. Smith (ed.s): *Early Modern Zoology. The Construction of Animals in Science, Literature and the Visual Arts.* vol. 2. Leiden & Boston: Brill, 2007, pp. 603–39. My text follows Sullivan's account. On Friedrich II and traditional falconry: Fans & Ritzau (2007).

Buffon: editions listed in notes to Chapter Five; 'indolent and melancholic' 7,247.

On gastronomy: Marx Rumpolt: *Ein new Kochbuch.* Frankfurt a.M.: Feyerabend, 1581. See also: J.H. Gurney: *Early Annals of Ornithology.* London: Witherby, 1921. Mrs Beaton's *Book of Household Management*, apart from discussing taste, associates the indiscriminate slaughter of wild birds for the table with the French. Her view of the slaughtering of birds such as the bittern stands next to her admiration for hunting, which she recommends, 'not only for the promotion of health but for helping to form that manliness of character which enters so largely into the composition of the sons of the British soil'. Johann Friedrich Blumenbach: *Handbuch der Naturgeschichte.* Göttingen: Dieterich, 1779.

Benjamin's celebrated account of aura in: *The Work of Art in the Age of Mechanical Reproduction.* Transl. Harry Zohn. New York: Schocken Books, 1969. His flippant comment on game butchers in: Short History of Photography, in: *Selected Writings*, vol. 2. Transl. Rodney Livingstone et al. Cambridge Ma. & London: Belknap, 1999, p. 522.

The colourful lives of the poachers Georg Jennerwein and Johann Adam Hasenstab need to be balanced out with the tragic story told by Theodor Fontane in the novel *Quitt*, in which Lehnert Menz is driven into exile when he kills a gamekeeper while poaching.

On Snyders: Kenneth Clark: *Animals and Men. Their Relationship as reflected in Western Art from prehistory to the present day.* London: Thames & Hudson, 1977, p. 30.

Hans Rudolf Schinz: *Naturgeschichte und Abbildungen der Menschen und der Säugethiere nach den neuesten Systemen und vorzüglichen Originalien bearbeitet.* Zürich: Honegger, 1827. vol. 2, p. 19.

Chapter Three

The full title of Frisch's work: *Vorstellung der Vögel Deutschlands und beyläufig auch einiger Fremden, nach ihren Eigenschaften beschrieben von Johann Leonhard Frisch [...] in Kupfer gebracht und nach ihren natürlichen Farben dargestellt von Ferdinand Helfreich Frisch.* Berlin: Birnstiel, 1763.

Pennant too had argued the need to get away from 'the chaos of Aldrovandi and Gesner' (Ritvo 1997, p. 17).

Quotation from Schinz (1827), vol. 2, p. 24. On the limitation of European species: Farber (1982), p. 76.

On Darwin and illustrations see: Voss (2009); & Jonathan Smith: *Charles Darwin and Victorian Visual Culture.* Cambridge & New York: CUP, 2006.

Seligmann's Edwards edition: *Sammlung verschiedener ausländischer und seltener Vögel [w]orinnen ein jeder derselben nicht nur auf das genaueste beschrieben, sondern auch in einer richtigen und sauber illustrirten Abbildung vorgestellt wird* von Johannes Michael Seligmann. Nuremberg: Johann Joseph Fleischmann, 1749. There are interesting comments on problems of nomenclature in the preface of the translator, Georg Leonhard Huth, MD. Edwards' remarks on colouring are in his preface (Neither text is paginated). A well-known example of the importance of colourists was when those working in Edinburgh on Audubon's work went on strike in 1827.

On the anonymity of colourists see: Christine E. Jackson & Maureen Lambourne, Bayfield – John Gould's unknown colourer, in: *Archives of Natural History*, 17 (1990), pp. 189–200. On Lear see: Jenny Uglow: *Edward Lear*. London: Faber & Faber, 2017. A better study will include both the wonderful artist Maria Sibylla *Merian and the engraver Johanna Dorothea *Sysang. Both have been in obscurity too long.

Jakob Theodor Klein: *Vorbereitung zu einer vollständigen Vögelhistorie.* Leipzig & Lübeck 1760.

The text describing the dissections carried out in France: *Der Herren Perrault, Charras und Dodarts Abhandlungen zur Naturgeschichte der Thiere und Pflanzen, welche ehemals der königlichen französischen Akademie der Wissenschaften vorgetragen worden, mit dazu gehörigen nach dem Leben gezeichneten Kupfern.* Aus dem Französischen übersetzt. Leipzig: Arkstee & Merkus, 1757.

There is a detailed account of the different modalities in bird painting as between skins, stuffed specimens etc (including the meaning of the phrase 'according to life') in Kinzelbach & Hölzinger (2000) (cf. notes to Chapter Two). Christine E. Jackson (1985) argues that the bird and branch style was initiated by Willughby and Ray, and reinforced by the taxidermists (pp. 27, 28).

Johann Karl Wilhelm Illiger: *Versuch einer systematischen vollständigen Terminologie für das Thierreich und Pflanzenreich*. Helmstädt: C.G. Fleckeisen, 1800, p. xxxii f.

Chapter Four

General works on Oudry: Mary Morton (ed.): *Oudry's Painted Menagerie. Portraits of Exotic Animals in Eighteenth-Century Europe*. Los Angeles: The J. Paul Getty Museum, 2007; and Hal N. Oppermann: *Jean-Baptiste Oudry*. New York & London: Garland, 1977. The first of these is the most richly illustrated, and the individual essays cover Oudry's relationship to other painters, including the animal painter Alexandre-François *Desportes.

On the cassowary see: Birkhead et al. (2014), p. 76.

On hunting see notes to Chapter Two.

Goethe praises Weenix extensively in *Dichtung und Wahrheit*.

On the menagerie see: Almudena Pérez de Tudela & Annemarie Jordan Gschwend, Renaissance Menageries. Exotic Animals and Pets at the Habsburg Courts in Iberia and Central Europe, in: Enekel & Smith (2007), vol. 2, pp. 419–47; Nigel Rothfels, How the Caged Bird Sings, in: Linda Kalof & Brigitte Resl (ed.s): *A Cultural History of Animals*. vol. 5: *In the Age of Empire*. Oxford & New York: Berg 2007, pp. 95–112; Maria Belozerskaya, Menageries as princely necessities, in: Morton (2007), pp. 58–73. On the relationship of the menagerie to other cultural values: Kurt Koenigsberger: *The Novel and the Menagerie. Totality, Englishness and Empire.* Columbus: Ohio State UP, 2007. On the 'camelopard' of the English king see: Isabella Tree: *The Bird Man: The Extraordinary Story of John Gould*. London: Ebury Press, 1991.

Buffon's position in pre-revolutionary France is discussed in: Wolf Lepenies: *Das Ende der Naturgeschichte. Der Wandel kultureller Selbstverständlichkeiten in den Wissenschaften des 18. und 19. Jahrhunderts.* Munich: Hanser, 1976. See also Chapter Five below. The dethroning of the lion king comes from Buffon's assistant Louis-Jean-Marie Daubenton (quoted Lovejoy (2015), p. 70 – details in notes to Chapter Five).

Michel Foucault: *The Order of Things* (1966). Translated 1970. Compare Loveland's summary: 'the mechanism of the early eighteenth century was replaced by a new vitalistic philosophy' (2015, p. 102).

For a critical view of Foucault's (and Lepenies') similar interpretation of the paradigm shift in natural history see: Farber (1982), pp. 124, 127f. On efforts to overcome anthropocentrism in the twentieth century see: Birkhead et al. (2014), pp. 245f.

Gustav Fechner: *Nanna, oder das Seelenleben der Pflanzen.* Leipzig: Voss, 1848. A comprehensive discussion of these holistic natural philosophies in: Walter Gebhard: *'Der Zusammenhang der Dinge'. Weltgleichnis und Naturverklärung im Totalitätsbewusstsein des 19. Jahrhunderts.* Tübingen: Niemeyer, 1894, pp. 164–221.

On Buffon's relationship to Oudry's work see Morton (2007), p. 16f.

On Le Brun see: Charles Le Brun: *A Method to Learn to Design the Passions* (1698: translation of 1734) with an introduction by Alan T. McKenzie. Los Angeles: University of California Press, 1980. Generally: Wilhelm Schlink: *Ein Bild ist kein Tatsachenbericht. Le Bruns Akademierede von 1667 über Poussins 'Mannawunder'.* Freiburg i.B.: Rombach, 1996. The relaxation of the hierarchy of painting subjects is generally observed. In the German context see i.a.: Helmut

Bernt: *Eine Berliner Künstlerkarriere im 18. Jahrhundert. Daniel Nikolaus Chodowiecki: Vom Kaufmannslehrling zum Medienstar*. Graz: Grazer Universitätsverlag, 2013. Chodowiecki too was influenced by Le Brun's work. Oudry's picture of a fox can be seen in Morton (2007, p. 86).

The Swedish National Museum set up an exhibition on the scientific debates round the expression of passions in art and science, and the richly illustrated catalogue remains an inspirational read, covering Darwin, medicine and all the arts. See esp.: Karen Sidén, Passioner. Konst och känslor genom fem sekler, in: *Passioner*: Stockholm, Nationalmuseum, 2012, pp. 10–35. An account of contemporary animal artists esp. Desportes and Stubbs in: Kenneth Clark (1977), p. 342.

A reference to the link to physiognomy in: Harriet Ritvo: *Darwin's Influence on Freud.* New Haven: Yale UP, 1991, p. 174f. Also: Ridley: *Darwin Becomes Art*. Amsterdam: Rodopi 2014, pp. 176–82. The central German figure is Carl Gustav Carus (1789–1869) and his study: *Symbolik der menschlichen Gestalt* (1852).

My account of American *exceptionalism* is indebted to John Updike's brilliant essay: The Clarity of Things, in: *NYRB*, 26 June 2008, pp. 12–16. Copley comes into Updike's argument in: *Still Looking. Essays on American Art.* Knopf: New York, 2006, pp. 11–26. Generally on Audubon: John Chancellor: *Audubon. A biography*. London: Viking Press, 1978; also Barbara Novak: *American Painting of Nineteenth-Century Realism.* New York: Harper & Row, 1979 (Chateaubriand quotation, p. 38). The internet is a cheaper route to find his pictures than the sale-room.

Chapter Five

Histoire Naturelle, générale et particulière, avec la description du Cabinet du Roi Paris, 1749–1804. The two Buffon editions on which the chapter is based are: *Allgemeine Historie der Natur. Aus dem Französischen übersetzt, mit Anmerkungen, Zusätzen und vielen Kupfern vermehrt durch* Friedrich Heinrich Wilhelm Martini. Berlin: Pauli, 1772f; and: *Allgemeine Historie der Natur nach allen ihren besonderen Theilen abgehandelt. Aus dem französischen mit Anmerkungen und Zusätzen.* Leipzig: Hermann Heinrich Holle, 1775–82 (Translator: Oehme). The third translation (there may well have been others, let alone anthology entries): *Herrn von Buffons Allgemeine Naturgeschichte*, transl. Joseph Georg Trassler. Troppau, 1785. An excellent account of Buffon in

English is: Jeff Loveland: *Rhetoric and Natural History. Buffon in polemical and literary context.* Oxford: Voltaire Foundation, 2015. Always useful are: Ritvo (1997), and Lepenies (1976) (see notes to Chapter Four). The English translation consulted is: *The Natural History of Birds. From the French of the Count de Buffon, illustrated with Engravings and a Preface, notes, and Additions, by the Translator.* e-book Cambridge: CUP. Place of publication not identified: publisher not identified, 1793. The title page identifies the translator as Mr Smellie, but the translator's preface makes clear that Smellie was not the translator, at least not of the whole work. Loveland stresses Smellie's rejection of Buffon's religious position: Buffon being 'a scandalous freethinker' (2015, p. 53). For convenience page numbers of quotations are given to this edition.

For an account of the controversy between Virchow and Häckel and the resulting politicisation of science see: Ridley (2014), pp. 44–61.

J.M. Bechstein: *Gemeinnützige Naturgeschichte Deutschlands nach allen drey Reichen. Ein Handbuch zur deutlichern und vollständigern Selbstbelehrung besonders für Forstmänner, Jugendlehrer und Oekonomen.* Leipzig: Crusius, 1791. Before writing his book Bechstein had translated both Levaillant and Latham. On Bechstein: *Johann Matthäus Bechstein (1757–1822) in den beruflichen und privaten Netzwerken seiner Zeit,* ed. Johannes Mötsch & Walter Uloth. Kassel: Norbert, 2009.

An excellent account of Linné's early reception in Germany: Ann-Mari Jönsson, The reception of Linnaeus' works in Germany, with particular reference to his conflict with Siegesbeck, in: *Germania Latina,* vol. 2, ed. E.Kessler & H.C.Kuhn. Munich: Fink, 2003, pp. 721–39. Buffon, as Loveland points out, also had his problems with the *Académie*. In his poem 'The Botanic Garden' Erasmus *Darwin refers to the reproductive organs of plants as their 'love organs'. On Erasmus Darwin see: Jenny Uglow: *The Lunar Men: The Friends Who Made the Future, 1730–1810.* London: Faber & Faber, 2002. Classification by means of reproductive organs was general in eighteenth-century France.

On holistic natural philosophy see: Gebhard (1984). (See notes to Chapter Four.) For a view of the scientific literature held in the sort of library Goethe and Oken fell out over, the Herzog August library in Wolfenbüttel, see: Petra Feuerstein-Herz: *Die große Kette der Wesen. Ordnungen in der Naturgeschichte der frühen Neuzeit.* Wolfenbüttel: Herzog August Bibliothek, 2007. Here too an account of the *scala natura,* the ancient concept revived in the fifteenth century. Its central statement: *natura non fecit saltus* (nature does not make jumps). This study adds to the classic account: Arthur O. Lovejoy: *The Great Chain of Being*. Cambridge Ma.: Harvard UP, 1936. See also Loveland (2015), pp. 77–99.

An outstanding study of the advance of science in nineteenth-century Germany: Andreas W. Daum: *Wissenschaftspopularisierung im 19. Jahrhundert. Bürgerliche Kultur, naturwissenschaftliche Bildung und die deutsche Öffentlichkeit 1848–1914.* Munich: Oldenbourg, 1998. Loveland's account of Buffon's 'vulgarisation' of natural history in: (2015), p. 36.

See: Oehme, Plan des ganzen Werks 3/16, p. 5. It's particularly striking that Oehme discusses at some length and positively Gesner's *Icones Animalium* (1553), even though that book was illustrated with wood-cuts and included pictures of animals Gesner had never seen (for instance, the rhinoceros) and of animals he could never see, such as the unicorn. Gesner clearly occupied a place in Natural History, even two centuries after his death. Buffon was the watershed for the survival of such books.

Translator's Preface: *loc. cit.* pp. iii-xi.

Johannes Michael Seligmann: See notes to Chapter Three.

Details of the translation of the post mortem records in notes to Chapter Four.

On Chodowiecki: Bernt (2013: See notes to Chapter Three). In 1864, when *Die Gartenlaube* published an obituary of the 'bird pastor' Christian Ludwig Brehm, Chodowiecki's copperplate resurfaced to accompany that obituary. In a characteristic sharp practice, it was described as being 'after the watercolour by Karl Werner in the British Museum'. The reference to the British Museum is invented. Neither the Print Room of the B. M., nor the archive of the Natural History Museum London (to which the scientific collections of the British Museum were transferred in the nineteenth century) contain any trace of it. It was Chodowiecki's original pirated and re-engraved.

The countryman's approach is epitomised in: Johann Andreas Naumann. Details of his books in notes to Chapter Seven. The link of hunting and observing is clear. When photography began to make inroads into the representation of nature, its similarity to hunting was a prime selling-point. Many photographic expressions came directly from the hunt, starting with 'shooting' and 'snap shot'. Liljefors was anything but unique in speaking of 'hunting with the camera', and the idea has continued into the present.

Naumann's photograph with his shot-gun is now in the museum at Köthen. The gun is genuinely enormous.

Classification problems are discussed in the following chapter. Oehme avoids the sort of argument which puts dogs and horses in the same phylum 'because they both follow the plough', but he gets tied up with the issues of 'land' or 'water' birds, which Aristotle had left behind.

Merrem's remark in: *Beitraege zur Naturgeschichte.* Duisburg & Lemgo 1790, p. 5. A glance at the internet will show that Merrem's obscurity is gradually lifting.

Quotations from On the Nature of Birds: 1,1–35. Quotation on method: 1, 2. 'Birds of prey [...] ferocious': 1, 41. Their tyranny 1, 38.

Tenderness as 'half envious', see: John Berger's essay: Why look at animals? in: *About Looking.* London: Viking, 1980, p. 11.

The Mastersingers of Nuremberg accused Walter of learning to sing from 'finches and tits'. For Wagner himself that accusation appears more as a compliment. The affinities in hearing between birds and humans comes down, according to a recent authority, to similarities in the structure of the inner ear. That was not something Buffon knew much about. See: Josef H. Reichholf: *Der Ursprung der Schönheit. Darwins großes Dilemma.* Munich: Beck, 2011, p. 219.

Oehme explains that Buffon holds to the view of reproduction as 'epigenesis', in which 'two types of seed' are involved, and the embryo develops from an undifferentiated (i.e., not pre-formed) cell. Oehme sees drawbacks with this theory, but gives it scientific weight. On this widely held theory see Loveland (2015), p. 58f.

Birdsong has tempted composers to imitation for centuries. In seventeenth-century Germany Athanasius Kircher explored both the expression of bird-song in musical notation and the links between anatomy and music. For Naumann bird-song provided evidence of the uses to which song-birds' larger brains were put, since birdsong requires memory (for extended melodies) and the ability to imitate. Naumann follows Buffon in seeing in the ability to imitate a predisposition which leads to an 'ease of taming' (Naumann & Buhle: *Die Vögel von Mittel-Europa und ihre Eier.* 3rd edition. Magdeburg: Creuzsche Verlagsbuchhandlung, n.d., p. 3.)

The topic of the sterility of hybrids is discussed in: Ritvo (1991), pp. 103–09.

Buffon's view of the Anthropocene (1, 13) is echoed in a striking feature of Alice Robert's recent study of domestication. The author regards entirely positively that 'humanising' of nature which Buffon identified in the quoted passage. She sees in the domesticated species a 'potential to be tame' and a 'predisposition on both sides'. *Tamed. Ten Species that Changed Our World.* London: Windmill, 2018, pp. 4, 21. Edwards was unusual in not regarding birds as being adaptable (loc. cit. p. 1f).

Schlegel's essay: Zweck und Eigenschaften naturkundlicher Abbildungen (transl. by Sophie Nissen) in: Claus Nissen (1978 –see notes to Preface), pp. 231–34, 338f.

On the cassowary 1,382.

Belon's bittern picture: Pierre Belon du Mans: *L'Histoire de la Nature des Oyseaux* (1555). Facsimile-edition, ed. Philippe Glardon. Geneva: Droz, 1977, p. 192. Lambourne too draws attention to another misleading shadow in Belon's work, this time in the picture of an avocet (1990, p. 102).

A standard work on Social Darwinism: Richard Hofstadter: *Social Darwinism in American Thought, 1860–1915.* Pennsylvania: University of Pennsylvania Press, 1944. The range of attitudes is enormous. Oehme would have found much to agree with. On Malthus: Robert M. Young: *Darwin's Metaphor. Nature's Place in Victorian Culture.* Cambridge: CUP, 1985, pp. 23–55.

Brown and black eagles 1, 55.

On the death conversation in falconry: Helen Macdonald: *H is for Hawk.* London: Jonathan Cape, 2014.

Chapter Six

There is a considerable literature on taxonomy. For taxonomy's place in ornithology Stresemann's history is excellent. So too: Birkhead et al. (2014, see notes to Preface). On a broader and more relaxed note see: Ritvo (1997).

Blasius Merrem's individual studies: *Beyträge zur besonderen Geschichte der Vögel.* Göttingen: J.F. Müllerschen Buchhandlung, 1784. General remarks in: *Beytraege zur Naturgeschichte.* Duisburg & Lemgo: im Selbstverlag, 1790, p. 5. August Johann Georg Carl Batsch: *Versuch einer Anleitung, zur Kenntnis und Geschichte der Thiere und Mineralien für akademische Vorlesungen entworfen, und mit den nötigen Bildern versehen.* Jena: Akademische

Buchhandlung, 1788. On 'fudging' in French classification see: Loveland (2015), p. 174.

Naumann's category included 'thick-beaked birds taking their food in trees'. Oken went in for 'Vogelvogel' and 'Aukauken'.

For a historical view of terminology see: Hermann Krause: *Die Geschichte der neueren zoologischen Nomenklatur in deutscher Sprache.* Inaug. Diss. Göttingen. Göttingen: Appelhans, 1918. Auch Suolahti: *Die deutschen Vogelnamen.* Strassburg: Trübner, 1909. Johann Karl Wilhelm Illiger: See notes to Chapter Three. Also Ritvo (1997), pp. 25, 62.

On Oken I have consulted: *Gesammelte Schriften*, ed. J. Schuster. Berlin: Keiper, 1939; *Elements of Physiophilosophy* (transl, by Alfred Turk). London: Ray Society, 1847; *Naturgeschichte für Schulen.* Leipzig: Brockhaus, 1821; & *Allgemeine Naturgeschichte.* Stuttgart: Hoffmann'sche Verlagsbuchhandlung, 1837. There are useful essays in: *Lorenz Oken 1779–1851. Ein politischer Naturphilosoph*, ed. Olaf Breidbach, Hans-Joachim Fliedner & Klaus Ries. Weimar: Böhlau, 2001. Here Thomas Bach's essay: 'Was ist das Thierreich anders als der anatomirte Mensch...?' (pp. 73–91). On Oken's foreign reception see: J.B. Stallo: *General Principles of the Philosophy of Nature with an outline of some of its recent developments among the Germans.* Boston: Crosby & Nichols, 1848. Also: Nicholas Jardine: Naturphilosophie and the kingdom of nature, in: Jardine, Secord & Spary (1996: see notes to Chapter One), pp. 230–45. For Oken's personal and scientific circle: Jean Strohl: *Lorenz Oken und Georg Büchner. Zwei Gestalten aus der Übergangszeit von Naturphilosophie zu Naturwissenschaft.* Zürich: Verlag der Corona, 1936.

Ludwig Brehm: *Lehrbuch der Naturgeschichte aller europäischen Vögel.* Part 1. Jena: August Schmid, 1823.

On the *Ignoramus* of university president du Bois-Raymond see: Robert J. Richards: *The Tragic Sense of Life. Ernst Haeckel and the Struggle over Evolutionary Thought.* Chicago & London: University of Chicago Press, 2008, p. 325.

On Swainson: my account bases on three texts: *Zoological Illustrations.* London: private printing, 1820f; *Treatise on Classification.* London: Longman, Rees et al., 1835; und *The Natural History and Classification of Birds.* London: Longman, Rees et al., 1836f. On the history of analogy see Foucault (1966). Notice the great increase in interest in fossils following Charles Lyell's *Principles of Geology* (1830f). The first bird fossil was not discovered until

1860. Ritvo (1997, p. 31f) gives a brief account of the Quinary system, and quotes Darwin's 'vicious circles' (p. 33).

On Chambers: Klaus Stiersdorfer, Vestiges of English Literature: Robert Chambers, in: Anne-Julia Zwierlein (ed.): *Unmapped Countries. Biological Visions in Nineteenth-Century Literature and Culture*. London: Anthem Press, 2005, pp. 27–30.

The place of mathematics in natural history was hotly debated in eighteenth-century France (see: Loveland (2015), pp. 127–52). On Oken and mathematics: Johanna Bohley, Gemeinsame Interessen […], in: Breidbach et al. (2001), p. 194f.

For Swainson's biblical starting-point see: Epistle of Paul to the Romans: i:20.

A discussion of the 'entangled bank' in: James Krasner: *The Entangled Eye. Visual perception and the representation of Nature in post-Darwinian Narrative*. New York & Oxford: OUP, 1992, pp. 33f, 173. Häckel and Jacobsen are discussed in Richards (2008) and Jensen (2017),

Chapter Seven

On Naumann's biography: Peter Thomson: *Johann Friedrich Naumann, der Altmeister der deutschen Vogelkunde. Sein Leben und seine Werke*. Revised by Erwin Stresemann. Leipzig: Barth, 1957. See also Stresemann's edition of the correspondence between Naumann and Schinz: *Die ornithologische Korrespondenz zwischen Johann Friedrich Naumann und Heinrich Rudolf Schinz in den Jahren 1815 bis 1835.* ed. Erwin Stresemann & Ludwig Baege. Odense: Odense Universitetsforlag, 1969. An interesting comparison with Alfred Brehm's career in: Hans-Dietrich Maemmerlein: *Der Sohn des Vogelpastors. Szenen, Bilder, Dokumente aus dem Leben von Alfred Edmund Brehm*. Berlin: Evangelische Verlagsanstalt, 1985. Brehm enjoyed close links with the important biologist and populariser of science, Emil Adolf Rossmässler (1806–67). See Daum (1998), p. 324.

Köthen is not even in Brandenburg: it's in Saxony/Anhalt.

Johann Andreas Naumann's practical hand-books: *Der Vogelsteller oder die Kunst allerley Arten von Vögeln sowohl ohne als auf dem Vogelherd bequem und in Mengen zu fangen.* Leipzig: Schwickert, 1789. See also a recent edition of selections: *Meistens selbst erfundene Fallen und Fänge zum Vogelstellen*. ed.

Ludwig Baege & Joachim Neumann. Köthen: Naumann-Museum, 1989. His field-guide in two volumes: *Naturgeschichte der Land und Wasservögel Deutschlands und angränzender Länder, nach eigenen Erfahrungen entworfen* [...] Köthen: Private printing, 1797f.

Johann Friedrich's twelve-volume edition: J.A. Naumanns *Naturgeschichte der Vögel Deutschlands, nach eigenen Erfahrungen entworfen. Durchaus umgearbeitet, [...] sehr vermehrt [...] und [...] aufs Neue herausgegeben* von [...] J. F. Naumann. (Fortsetzung der Nachträge, Zusätze und Verbesserungen von Dr. J. H. Blasius, Dr. E. Baldamus und Dr. F. Sturm.) Leipzig: Fleischer, 1822f.

Bewick also used the category of 'usefulness', see: Uglow (2006), p. 252f.

On the *Zugunruhe* see Birkhead et al. (2014), p.143.

My comments on nineteenth-century biology in Germany are indebted to the superb work of Lynn K. Nyhart, Natural History and the New Biology, in: Jardine, Secord and Spary (1996), pp. 426–43; & *Modern Nature. The Rise of the Biological Perspective in Germany.* Chicago & London: University of Chicago Press, 2009. Quotations are from the latter, pp. 4 & 18. On popularisation of science see: Nyhart (1996, p. 443) and Daum (1998).

Darwin letter of 1861 in: Alvar Ellegård: *Darwin and the General Reader. The Reception of Darwin's Theory of Evolution in the British Periodical Press 1859–1872.* Göteborg: Goteborg Studies in English vii, 1958, p. 194.

The idea of reading the body as a language, with an alphabet, lexis and grammar was familiar among the physiognomists and Darwin shows familiarity with the metaphor. See Ridley (2014), p. 188f. Ralph Waldo Emerson also uses the metaphor in the essay 'Beauty'.

Balzac's *L'Épicier*, in: *Les Français peints par eux-mêmes.* vol. 1. Paris: Curmer, 1840. See the brilliant study by Martina Lauster: *Sketches of the Nineteenth Century. European Journalism and its 'Physiologies'.* Houndmills: Palgrave, 2007, p. 94f.

Darwin and Wallace were particularly interested in the function of coloration in birds, for instance as between male and female. Rather than thinking of male birds' bright colours as an expression of their 'character' (we see Darwin's problems with this cliché in the next chapter), function takes over. See the fuller notes to Chapter Ten on the zoological reading of colour.

'A discourse through which nature and society could be discussed' (Nyhart 2009, p. 4) echoes Young (1985), p. 238: 'the point about evolutionary theory is that it is the central conception linking humanity and social theory to natural science'.

The phrase 'high-brow' comes directly from the Dutch ethnologist Peter Camper. See: Kemp (2003), p. 129f and Ritvo (1997), p. 121f.

Details of Schlegel's essay in notes to Chapter Five.

On Gould as an 'invisible helper' see: Tree (1991).

Gessner's view of illustration emerges from the account of a gifted eighteenth-century Swiss natural history artist, specialized in insects: Brigitte Thanner, Hans-Konrad Schmitz und Arnim Geus: *Johann Rudolf Schellenberg. Der Künstler und die naturwissenschaftliche Illustration im 18. Jahrhundert*. Winterthur: Stadtbibliothek, 1987, pp. 63f, 142f.

The entomologist's comments on illustrations: Theodor Wohlfahrt, Schmetterlinge in der Illustration, in: Nissen (1978), pp. 306–326.

Naumann's defeated rivals: Bernhard Meyer und Johann Wolf: *Naturgeschichte der Vögel Deutschlands in getreuen Abbildungen und Beschreibungen.* Nuremberg: Kunstverlag Frauenholz, 1805f. On the shift from decorative and theological painting of birds see: Heidrun Ludwig: *Nürnberger naturgeschichtliche Malerei im 17. und 18. Jahrhundert.* Marburg: Basilisken, 1998, p. 180f. Notice too, as Jenny Uglow (2017) sensitively establishes, how Edward Lear found a way out of bird painting (both as slavery to Gould and as a demanding, yet not prestigious discipline) via landscape painting (p. 98).

Chapter Eight

The best English language source to Wolf's life and works is: Alfred H. Palmer: *Life of Joseph Wolf. Animal Painter*. London & New York: Longmans, Green & Co, 1895, which includes a complete list of Wolf's pictures. A good biographical account and a wide range of his bird portraits can be found in: S. Peter Dance: *Birds of Prey by Josef Wolf*. London: Studio Editions, 1991. Dance sees Wolf as being 'without exception the best all-round animal painter that ever lived' (p. 5). The more complete source is: *Josef Wolf (1820–1899). Tiermaler / Animal Painter.* ed. Karl Schulze-Hagen & Arnim Geus. Marburg: Basilisken, 2000. Here Christine E. Jackson writes of Wolf as a 'turning-point' in animal painting (pp. 173–88). Also important essays on Wolf's collaboration with

Daniel Giraud Elliot on the birds of North America: David M. Lank, Wolfs naturkundliche Illustrationen, (pp. 143–71).

On the history of biology in nineteenth-century Germany see: Nyhart (1996 & 2009). On Kaup: Ernst Probst: *Johann Jakob Kaup. Der große Naturforscher aus Darmstadt. 1803–1873.* Munich: GRIN, 2011.

Charles Darwin: *The Expression of the Emotions in Man and Animals* (1872), transl. by J. Victor Carus (1877). Voss (cf. notes to Chapter One) gives an account of Wolf's involvement other public debates including Du Chaillu's discovery of the gorilla species. For Darwin's relationship to his illustrators see: Smith (2006) (cf. notes to Chapter One).

On the early history of ethology and the shifts brought about by Tinbergen and Wilson: W.H. Thorpe: *The Origins and Rise of Ethology.* London: Heinemann, 1979. Thorpe acknowledges the importance of bird illustration in the early history of the subject. It was the illustrators who, confronted for instance with the skins of the bird of paradise – dispatched to Europe often without wings or legs – had to decide and then portray how the species actually behaved in nature

On George Eliot and science see: Sally Shuttleworth: *George Eliot and Nineteenth-Century Science. The Make-Believe of a Beginning.* Cambridge: CUP, 1984.

The early phases of Darwin reception are described by: Peter Bowler: *The Eclipse of Darwin. Anti-Darwinian Evolution Theories in the Decades around 1900.* Baltimore & London: John Hopkins UP, 1983. Also: Thomas F. Glick und Eve-Marie Engels (ed.s): *The Reception of Charles Darwin in Europe.* London: Continuum, 2008. Also Thomas F. Glick & Elinor Shaffer (ed.s): *The Comparative Reception of Darwinism.* London: Bloomsbury, 2014.

The phrase 'pascha of a harem' was used by: Adolf & Karl Müller: *Thiere der Heimath. Deutschlands Säugetiere und Vögel. Mit Original-Illustrationen nach Zeichnungen auf Holz und Stein von C.F. Deiker und Adolf Müller.* Kassel und Berlin: Th. Fischer, 1882. (In this volume there is one of Deiker's wild boar pictures (cf. Chapter Ten).

Wolf's *Nosey Neighbours* is reproduced in: Palmer (1895), p. 228.

Heidrun Ludwig: see notes to Chapter Seven.

Nissen's remarks are in his volume: *Die illüstrierten Vogelbücher* (1953), p. 23f.

Liljefors got to know Wolf's pictures through: Llewellyn Lloyd: *The Game Birds and Wild Fowl of Sweden.* London: Warne, 1867. Wolf's woodcock is examined by: Maureen Lambourne, John Gould and Josef Wolf, in: Schulze-Hagen & Geus (2000), pp. 173–88. My text also mentions the two pictures *Beckasin* and *Ejdersträck*: both hang in the Thielska Galleriet Stockholm and are reproduced in: *Ute I markerna* (cf. notes to Chapter Ten).

Chapter Nine

On the general history of photography: Helmut & Alison Gernsheim: *A Concise History of Photography.* London: Thames & Hudson, 1965. Gernsheim uses the word realism somewhat uncritically. Much more reflective is the essay of Walter Benjamin, Short History of Photography, (1931) (see notes to Chapter Two). See also: Jonathan Crary: *Techniques of the Observer. On Vision and Modernity in the Nineteenth Century.* Cambridge, Ma. & London: MIT Press, 1996. On technical questions: Frank Heidtmann: *Wie das Photo ins Buch kam. Der Weg zum photographisch illustrierten Buch [...]* Berlin: Spitz, 1984. The quotation about portraits comes from *Photographic News* (London, 1861) quoted by Gernsheim, p. 119. His chapter on portrait photography is strongly recommended. Reading further in the early numbers of the *Journal of the Photographic Society* (London, 1854f) one finds almost word for word Meerwarth's argument concerning detail in the photo. Arguments against photography at the time used the same vocabulary as Meerwarth.

H. Meerwarth (ed.): *Lebensbilder aus der Tierwelt. 2. Reihe: Vögel 1.* Leipzig: Voigtländer, n.d. (1906). Also: *Photographische Naturstudien. Eine Anleitung für Amateure und Naturfreunde.* Erlangen & Munich: Schreiber, 1905.

Compare: A. Radclyffe Dugmore: *Wild Life and the Camera.* London: Heinemann, 1912, esp. the first chapter: Bird Photography, p. 3f.

Ernst Häckel: *Kunstformen der Natur.* Leipzig: Bibliographisches Institut, 1904. Benjamin mentions another volume of similar close-ups: Karl Blossfeldt: *Urformen der Kunst. Photographische Pflanzenbilder von Professor K.B.* Berlin: Wasmuth, 1928.

A contemporary view of colour photography: Arthur Freiherr von Hübl: *Die Dreifarbenphotographie mit besonderer Berücksichtigung des Dreifarbendruckes und ähnlicher Verfahren.* Halle: Knapp, 1912/3. Hübl had been known since the 1880s as the inventor of platinum printing.

On Strindberg's understanding of photography: August Strindberg: *Notizen eines Zweiflers. Schriften aus dem Nachlass*, transl. Renate Bleibtreu. Berlin: Berenberg, 2011. See further: George E.F. Schulz: *Natururkunden. Biologisch erläuterte Photographien freilebender Tiere und Pflanzen. Hf.1: Vögel*. Berlin: Parey, 1908. Also: P.H. Emerson: *Naturalistic photography for students of the art*. London: S. Low, Marston, Searle & Rivington, 1889.

George E. Lodge: *Memoirs of an Artist-Naturalist*. Edinburgh: Gurney & Jackson, 1946.

Pierre Bourdieu: *Un art moyen. Essai sur les usages sociaux de la photographie*. Paris: Éditions de minuit, 1965; John Berger (1972) (see notes to Chapter Five).

Baudelaire's rejection of photography in: Le public moderne et la photographie, in: *Le Salon de 1859. Oeuvres Complètes. Curiosités Esthétiques*. Paris: Conard, 1923, pp. 264–72.

Naumann's egg-book: Johann Friedrich Naumann &. Dr. Christian Adolph Buhle: *Die Eier der Vögel Deutschlands und der benachbarten Länder*. Halle: Kümmel, 1818.

The great names from photojournalism in war were: Roger Fenton (Crimean War), Mathew Brady, Timothy O'Sullivan und Alexander Gardner (US-Civil War) (Photography as 'the eye of history' quoted in: Gernsheim, p. 142).

Ledwig Davis: *Ratgeber im Photographieren. Für Anfänger und Fortgeschrittene*. Halle: Knapp, 1912. Jean Anker (See notes to Preface).

On Illiger's concept of Habitus see Chapter Three. Kemp (1990) (See notes to Chapter One).

Chapter Ten

Svetlana Alpers: see notes to Chapter One. Reinhard Piper: *Das Tier in der Kunst*. Munich: Piper, 1912.

Works by Liljefors: *In the realm of the wild: the art of Bruno Liljefors of Sweden (1937)*. Göteborg: Göteborgs konstmuseum, 1988. See the representative selection from his work: *Ute i Markerna. Reproduktioner efter tavlor av B.L.* Stockholm: Bonniers, n.d. The most complete biography is: Allan Ellenius. *Bruno Liljefors*. Uppsala: Carmina, 1981. On his work and its reception see: Jacques Tersmeden: *Bruno Liljefors. Konst och person inför svensk och utländsk allmänhet 1890–1897*. Umeå: Umeå Universitet, 1985. Liljefors' book of photos, together with Paul Rosenius: *Naturstycken, naturskildingar.* Stockholm, Wahlström & Widstrand, 1896. On Liljefors' relationship to photography: Karin Sidén & Göran Söderlund: *Ett med naturen. Bruno Liljefors och Naturfotograferna.* Stockholm: Carlsson Bokförlag, 2013. On Liljefors' place in Swedish art-history: *Signums svenska konsthistoria. Konsten 1890–1915.* Lund: 2001, pp. 221–25. Interesting comments by the present-day master of bird painting, Lars Jonsson, on his great model Liljefors in: *Where Heaven and Earth Touch: The Art of the Birdpainter Lars Jonsson.* Petersberg: Michael Imhof, 2008.

A wonderful source-book on the relationship between Scandinavian and German artistic life in the nineteenth century is the exhibition catalogue produced by the Deutsches Historisches Museum, the Norsk Folkemuseum and the Swedish National Museum: *Wahlverwandtschaft. Skandinavien und Deutschland 1800 bis 1914.* ed. Bernd Henningsen, Janine Klein, Helmut Müssener & Solfrid Söderlind. Here for instance an essay on the links in Impressionism: Cecilia Lengefeld, Max Liebermann und Anders Zorn, pp. 358–62.

General information on Deiker from: Ulrich Thieme & Felix Becker (ed.s): *Allgemeines Lexikon der bildenden Künstler. Von der Antike bis zur Gegenwart.* Leipzig: Engelmann, 1907f.

The comparative exhibition on Nolde and Liebermann: Martin Faass (ed.): *Max Liebermann und Emil Nolde: Gartenbilder; Katalogbuch zur Ausstellung in Berlin-Wannsee, Liebermann-Villa am Wannsee, 22.04.–20.08.2012.* Berlin 2012.

George Eliot takes her heroine Dorothea into that circle on her honeymoon (see: *Middlemarch*). Edward Lear paid many visits to Rome in this fashion.

Jacobsen's *Niels Lyhne* was immediately translated into German. An English translation by Tina Nunnally is available in Penguin Books. On the general context see: Klaus Bohnen (ed.): *Fin de siècle. Zu Naturwissenschaft und Literatur der Jahrhundertwende im deutsch-skandinavischen Kontext.* Kopenhagen & Munich: Fink, 1984. Benn's understanding of Jacobsen and Jacobsen's presentation of Darwinism are fully analysed in: Ridley *(*2014), p. 107f. This has been complemented by Jensen's recent biography (2017). Jensen stresses in particular Darwin's effect of 'deposing' man as the centre of the world (p. 41). This had major consequences not just on Jacobsen's religion, but on his perception of nature.

Despite Edschmid's emphatic clarification, there's a basic confusion in this cultural period concerning the actual relationship between Impressionism and Expressionism. In part, the problem arises because of the fact that for two of the major European poets of this period, Gottfried Benn and Hugo von Hofmannsthal, the two movements represented stages of their own personal development. Hofmannsthal admired Niels Lyhne's passivity, but a very few years later he's writing to Maria Herzfeld (the great advocate of Jacobsen in Germany): 'I'm thoroughly tired of all the refinement, subtlety, fine textured impressionistic, psychological stuff, and I can't wait for the naïve pleasures of life to drop – with the earthy smell of pine cones – from the trees.' Benn too wrote his early poetry in obvious affinity with the Impressionist Detlev von Liliencron, but shortly afterwards was identified as an apologist for Expressionism. In other words: between Impressionism and Expressionism there's a grey area, especially in poetry and short prose. It's for that reason that I quote Benn so often in these pages; from his Expressionist phase comes the remarkable essay on van Gogh: *Der Garten von Arles* (1920), and from a later period his summary of the 'asiatic' (i.e., 'static') art favoured by his generation. *Gespräch* appeared in the Impressionist phase of his work, but opened an awareness of the scientific dimension of Jacobsen's work which Benn could develop further in his Expressionist phase. *Gespräch* in: *Szenen und Schriften in der Fassung der Erstdrucke*, ed. Bruno Hillebrand. Frankfurt a.M.: Fischer, 1990, p. 13f. *Der Garten von Arles*, in: *Prosa und Autobiographie in der Fassung der Erstdrucke, ibid.* p. 91f. On the 'asiatic' see: 'Statische Metaphysik' in: *Der Roman des Phänotyp* (1944), *ibid.* p. 154f. Rilke's Roman – *Die Aufzeichnungen des Malte Laurids Brigge* (The Notebooks of Malte Laurids Brigge) – loosely based on Jacobsen, appeared in 1910 und established itself as a canonical text of modernism. Hofmannsthal's so-called *Chandos Letter* had appeared in 1902 and joined Rilke's novel in that canon.

On Liljefors and the scientific discussion of colours see: Ellenius (1981), p. 127f. It shows the general interest of the theme that a work on Wallace and colour appeared in 1890, written by a prominent sociologist, rather than biologist: Gustav Fredrik Steffen: *Färgernas Betydelse i Djur-och Växtvärlden. En Framställning Enligt Wallace [in His Work 'Darwinism']*. Nietzsche's reference to mimicry is part of his discussion of Darwin in: 'Anti-Darwin' in *Twilight of the Idols*. The standard work of that period: Frank Evers Beddard: *Animal*

Coloration: An Account of the Principal Facts and Theories Relating to the Colours and Markings of Animals. London: Swan Sonnenschein, 1892.

Grez-sur-Loing is in the neighbourhood of Fontainebleu. Various Swedish and American painters stayed here, among them: Carl Larsson, John Lavery and the composer Frederick Delius – all in honour of Impressionist plein air painting.

John Berger: Why look at animals?' (See notes to Chapter five.)

Material quoted on Impressionism is taken from a classic text of art-history: Richard Hamann & Jost Hermand: *Impressionismus*. Berlin: Akademie Vg., 1966. By the same authors: *Expressionismus* (*ibid*. 1975).

The figure of 'Great Pan' makes its appearance in Nietzsche's work, esp. in: *Thus Spake Zarathustra* (1884). A widely read novel of Hamsun's has the title *Pan* (1894).

On Schrödinger and the robin see: Jim Al-Khalili & Johnjoe McFadden, both in the *Guardian* of 14.12.14; and in: *Life on the Edge. The Coming of Age of Quantum Biology*. New York: Bantam, 2014.

On the influence of the art of Japan and China there is a huge range of literature. I consulted: Ernst H. Gombrich: *Die Geschichte der Kunst*. Stuttgart: Besler, 1992^5, p. 443f; Musée Guimet (ed.): *La tentation de l'Orient*. Paris: Réunion des musées nationaux, 1988. More work needs to be done on the impact of the import of decorative Chinese art in the eighteenth century. It is mentioned in connection with Oudry's work. In the summer palace of Queen Luise of Prussia in Paretz (Brandenburg) one can still admire the painted wallpaper, imported from China in 1795 and displaying a wonderful range of birds. Liljefors himself was not interested in Chinese art. For him it lacked a homespun quality and was 'precious' (cit. Ellenius (1981), p. 74).

Gould's colibri-book: John Gould: *An introduction to the Trochilidae, or family of humming-birds.* London: J. Gould, 1861. See Chapter Eight. On Heade see: Pamela Kort & Max Hollein (ed.s): *Darwin: Kunst und die Suche nach den Ursprüngen.* Köln: Wienand, 2009, p. 14f.

Magdalene M. Müller published two books on Nolde's expedition: *Emil Nolde. Expedition in die Südsee.* Munich: Hirmer, 2002, &: *Emil Nolde in der Südsee.* Berlin: Brücke Museum, 2002 (this volume includes the pictures from that journey). I wrote more extensively on this episode in: *'Relations stop nowhere'. The Common Literary Foundations of German and American Literature 1830–1917.* Amsterdam & New York: Rodopi, 2007, pp. 201–46. Melville and Gerstäcker are included in this discussion.

On vitalism: Gunter Martens: *Vitalismus und Expressionismus. Ein Beitrag zur Genese und Deutung expressionistischer Stilstrukturen und Motive.* Stuttgart: Kohlhammer, 1971. On the important history of the concept of the 'life force' in the natural sciences see: Timothy Lenoir: *The Strategy of Life. Teleology and Mechanics in Nineteenth-Century German Biology.* Dordrecht, Boston & London: Reidel, 1982. The German best-seller is Wilhelm Bölsche: *Das Liebesleben in der Natur. Eine Entwicklungsgeschichte der Liebe.* Jena: Diederichs, 1908. The idea of the life-force represented a significant affinity between science and philosophy, even after Helmholtz had debunked the 'vital force'.

The most famous discussion of African sculptures in Germany was: Carl Einstein: *Negerplastik*. Leipzig: Weiße Bücher, 1915. The classic account of the Expressionists' cult of 'primitivism' is: Eckart von Sydow: *Die deutsche expressionistische Kultur und Malerei.* Berlin: Furche, 1920.

Rousseau's picture hangs in der National Gallery Washington.

Palau is one of a group of islands in the Western Pacific, sold to Germany in 1899 – now, more happily, a presidential republic – visited the year before Nolde's journey by Max Petzold, another Expressionist painter.

Brief Biographical Notes

Aesop: A slave and storyteller believed to have lived in ancient Greece between 620 and 564 BCE. His fables have been adapted by numerous writers.

Louis **Agassiz** (1807–73): Born in Switzerland, studied in Germany, took up a professorship in natural history in Harvard in 1847. A prominent opponent of Darwin's theory of evolution, also on matters of ethnicity.

Eleazar **Albin** (1690–1742): Bird and insect painter who emigrated from Germany to England. His *A Natural History of Birds* (1731–38) is the first bird book with coloured illustrations.

Ulisse **Aldrovandi** (1522–1605): A celebrated Italian doctor and natural historian. His *Ornithologiae libri* appeared from 1599. Created the Bologna botanical garden.

Ottomar **Anschütz** (1846–1907): Chronophotographer, specialising in birds, but also in high-speed photographs of bullets and shells.

Joseph **Arnold**: Quoted by Frisch, but not identifiable.

Aristotle's *Historia Animalium* (History of Animals) is one of the major texts on zoology from antiquity.

John James (Jean-Jacques) **Audubon** (1785–1851): Pioneering ornithologist and bird painter in North America. His *Birds of America* appeared from 1827 to 1838. Belatedly there has been a negative response to his activities as a slave-holder.

Leonhard **Baldner** (1612–94): Strasbourg fisherman and naturalist who produced an illustrated manuscript of natural history.

Honoré de **Balzac** (1799–1850): Major French Realist novelist. Author of animal satires *Scènes de la vie privée et publique des animaux* (1840f).

Jacopo di **Barbari** (ca. 1460–1516): Italian painter who moved to Nuremberg in 1500 and worked with Dürer.

Francis **Barlow** (1626–1704): English painter and illustrator, known for the accuracy of his paintings of animals.

Jacques **Barraband** (1768–1809): French animal and botanical painter, known for his pictures of tropical birds. Illustrator for Le Vaillant's expeditions.

August **Batsch** (1761–1802): Botanist professor of Natural History, Medicine and Philosophy in Jena.

Johann Matthäus **Bechstein** (1757–1822): Ornithologist and director of the Forestry Academy in Dreißigacker (part of Meiningen, Thuringia).

Charles **Bell** (1774–1842): Anatomist in Edinburgh, specialised in anatomical illustration. His *Essays on The Anatomy of Expression in Painting* (1806) held diametrically opposed views to Darwin's *Expression of the Emotions*...

Pierre **Belon** (1517–64): French natural historian and author of the celebrated *L'histoire des Oiseaux* (1555).

Gottfried **Benn** (1886–1956): Doctor and lyric poet. Celebrated after 1945, despite his temporary support for Hitler.

Thomas **Bewick** (1753–1828): English naturalist and wood-cut artist.

Johann **Blumenbach** (1752–1840): Professor of medicine and natural history in Göttingen. Taught both Alexander von Humboldt and the English poet Samuel Coleridge. Elected a Foreign Member of the Royal Society of London in 1793.

Pieter **Boel** (1622–1674): Flemish painter and illustrator, best known for his work in France, especially animal painting.

Albert **Brehm** (1829–1884): Zoologist, director of zoological gardens and author of *Brehms Tierleben*. Son of Christian Ludwig Brehm (1787–1864), the 'bird pastor'.

Mathurin-Jacques **Brisson** (1723–1806): Curator and ornithologist, author of the attempt at taxonomy *Ornithologie* (1760). Cf. Farber (1982), p. 8.

Georges-Louis Leclerc, Comte de **Buffon** (1707–1788): French naturalist and polymath, director at the Jardin du Roi. Author of *Histoire naturelle, générale et particulière* (1749–88: in 36 volumes). **See Chapter Five.**

Petrus **Camper** (1722–89): Dutch medic and anatomist, specialised in skull shapes, including 'low brow' etc.

Mark **Catesby** (1683–1749): English naturalist specialising in flora and fauna of North America. Also known as bird painter.

Robert **Chambers** (1802–71): His highly popular *Vestiges of the Natural History of Creation* (1844) anticipated both evolutionary ideas and some speculative pseudo-science.

Daniel **Chodowiecki** (1726–1801): Celebrated painter and etcher in Berlin. Influential director of the Berliner Kunstakademie.

John Singleton **Copley** (1738–1815): Moved from Boston Ma. to London, where he tried to establish himself as a society portrait painter.

Georges **Cuvier** (1769–1832): French anatomist and palaeontologist, associated with catastrophe theory.

Erasmus **Darwin** (1731–1802): English naturalist, philosopher and poet. Grandfather of Charles Darwin.

Edmé-Louis **Daubenton** (1730–85): French illustrator, who supervised the coloured illustrations for Buffon's monumental *Histoire Naturelle* (1749–89).

Jacques-Louis **David** (1748–1825): French republican history painter, best known for *Napoleon at the St Bernard Pass* (1801).

Alexander-François **Desportes** (1661–1743): French painter of animals and the hunt, a pupil of Snyders.

Dollmer: Collector in California, consulted by Merrem. Regrettably no further details.

George **Edwards** (1694–1773): Widely travelled English ornithologist, worked for many years as librarian to the Royal College of Physicians in London

Daniel Giraud **Elliot** (1835–1915): American zoologist and the founder of the American Ornithologist Union. Used Smit and Wolf for his illustrations.

Jan van **Eyck** (1390–1441): Supported by a court patron, van Eyck painted a much wider range of themes and a wide range of countries.

Carel Pietersz. **Fabritius** (1622–54): Dutch painter, pupil of Rembrandt, famous for his portrait of the chained goldfinch.

Gustave **Flaubert** (1821–80): major French Realist novelist, best known for *Madame Bovary* (1857).

Georg **Forster** (1754–94): Explorer, naturalist and revolutionary; son of Johann Reinhold Forster (1729–98), a pastor and natural historian who accompanied James Cook on his second expedition.

Friedrich II (1194–1350): From 1198 King of Sicily and from 1220 Holy Roman Emperor. A Hohenstaufen.

Caspar David **Friedrich** (1774–1840): The most significant painter of the early Romantic movement. born in Greifswald. Blue is a marked feature of his painting *Monk by the Sea*.

Ferdinand Helfreich **Frisch** (1707–58): Illustrator and son of the lead author of an important handbook, Johann Leonhard **Frisch** (1666–1743), scholar, teacher, friend of Leibnitz. **See Chapter Three.**

Francis **Galton** (1822–1911): half-cousin to Charles Darwin, leading eugenics theorist, claiming to derive inspiration from Darwin's *Origin of Species*.

Conrad **Gesner** (1516–65): Swiss scholar and natural historian. Author of the widely received *Icones Avium* (1560).

Salomon **Gessner** (1730–88): Well-known Swiss painter and writer.

Johann Wolfgang v. **Goethe** (1749–1832): Hugely prolific writer and poet, author of scientific studies (notably on plants and on colours); for a while held high office in the Duchy of Weimar. Often seen as the national poet of Germany.

John **Gould** (1804–81): Celebrated ornithologist, cataloguer of Australian fauna. Close to Darwin. FRS.

Ernst **Häckel** (1834–1919): Marine biologist, enthusiastic and controversial supporter of Darwin, major influence on language of biology, famous for his biogenetic law.

Martin Johnson **Heade** (1819–1904): American painter, learned with Hicks, close links to the Hudson River School, settled in Florida.

Christian Johann Heinrich **Heine** (1797–1856): German poet and essayist, author of many poems set by Schumann and Schubert. Spent 25 years of his life in exile in Paris.

Johann Gottfried v. **Herder** (1744–1803): German Enlightenment figure renowned for his views on language and on different cultures.

Edward **Hicks** (1780–1849): American folk painter ('American Primitives') and Quaker pastor. Created iconic images of early USA life.

Ernst Theodor Amadeus **Hoffmann** (1776–1822): Prolific German writer of fantasy and Gothic novels: a major figure of European Romanticism.

Hugo Lorenz August Hofmann von **Hofmannsthal** (1874–1929): Austrian prodigy as lyric poet, dramatist and librettist (*Der Rosenkavalier*).

Arno **Holz** (1863–1929): German Naturalist dramatist, theoretician and poet.

Melchior **Houdecoeter** (1636–95): Highly gifted Dutch animal and bird painter.

Alexander v. **Humboldt** (1769–1859): German explorer, naturalist and philosopher of science. His most famous journey was from 1799–1804 round parts of Central and Southern America. Brother to Wilhelm von **Humboldt** (1767–1835), linguist, diplomat, minister of state and founder of the University of Berlin.

Johann Karl Wilhelm **Illiger** (1775–1813): Celebrated zoologist, director of what became Berlin's Natural History Museum.

Jean-Auguste-Dominique **Ingrès** (1780–867): French Neo-Classical painter in the tradition of Jacques-Louis David.

Jens Peter **Jacobsen** (1847–85): Danish botanist and novelist, translator of Darwin. Author of the novel *Niels Lynhe*.

Lars **Jonsson** (1952): Swedish ornithologist and bird artist. Considerable output, e.g., *Lars Jonsson's Birds* (2008).

Johann **Kaup** (1803–73): Researcher and natural historian in the Darmstadt Museum.

Johannes Gerardus **Keulemans** (1842–1912): Dutch painter was supported by Schlegel and worked mostly in London.

Jacob Theodor **Klein** (1685–1759): German diplomat and zoologist working on taxonomy in Danzig (Gdansk). Cf. Farber (1982), p. 4.

Pauline **Knip** (née Pauline Rifer de Courcelles): (1781–1851): French bird painter, pupil of Barraband, collaborated with Coenraad Temminck.

Jean de **la Fontaine** (1621–95): Highly popular French writer of fables.

Jean-Baptiste Pierre Antoine de Monet, chevalier de **Lamarck** (1744–1829): French evolutionary botanist.

Marcus zum **Lamm** (1544–1606): German ecclesiastical lawyer who produced a 33-volume illustrated work of ornithology, the *Thesaurus Picturarum*.

Edward **Landseer** (1802–73): Animal painter and sculptor, essential part of the Victorian establishment.

John **Latham** (1740–1837): English physician and ornithologist. *A General Synopsis of Birds* (1781f) marked by classification uncertainties.

Johann Kasper **Lavater** (1741–1801): Swiss pastor and founder of physiognomy. Set the fashion for silhouette portraits.

Charles **Le Brun** (1619–90): President of the Salon, court painter and director of the Gobelins factory.

François **Levaillant** (1753–1824): Explorer, collector & ornithologist, completing major journeys in various continents.

Bruno **Liljefors** (1860–1939): Swedish nature and bird painter. ***See Chapter Ten.***

Carl v. **Linné** (1707–78): Celebrated Swedish botanist and zoologist; his *Systema Natura*, esp. in 10th edition of 1758, established binomial nomenclature, the so-called Linnéan system,

Michael Matthias **Ludolff** (1685–1756): Professor of botany and medicine in Berlin.

William **MacGillivray** (1796–1852): Scottish naturalist and ornithologist. Professor in Aberdeen. Accomplished bird painter.

Thomas Robert **Malthus** 1766–1834): Economist and social philosopher, who warned in *An Essay on the Principle of Population* against overpopulation and impoverishment.

Étienne-Jules **Marey** (1830–1904): French physiologist and chronophotographer. *Le Vol des Oiseaux* (1890) contains the results of the work on birds' flight.

François-Nicholas **Martinet** (1731–1804): Illustrator and engraver engaged by Buffon for his illustrations.

Friedrich Heinrich **Martini** (1729–78): German physician, conchologist and translator i.a. of Buffon.

Meister der Spielkarten: one of the earliest master engravers, best known for his many playing-cards designs, among which there are some wonderful birds. Active in the second half of the fifteenth century.

Adolf **Menzel (**1815–1905): Major German Realist painter, strongly associated with Berlin. Many famous historical paintings, together with iconic industrial scenes, notably in the *Eisenwalzwerk* (1872).

Maria Sibylla **Merian** (1647–1717): Wonderful German insect painter and entomologist. Went on an important research stay in Surinam in 1699–71.

Blasius **Merrem** (1761–1825): Pupil of Blumenbach, subsequently professor in Duisburg and Marburg.

Frans van **Mieris** (1635–81): Dutch painter celebrated for his *Young Woman Feeding a Parrot.*

K.A. **Möbius** (1825–1908): Marine biologist, director of the Zoological Museum in Berlin.

Killian **Mullarney** (1958): Irish ornithologist and bird artist, responsible, with Dan Zetterström for illustrating the *Collins Bird Guide*, and *Birds of the Western Palearctic.*

Edvard **Munch** (1863–1944): Norwegian painter, best known for *The Scream*. Close to van Gogh, Gauguin, Strindberg and other modernists.

Johann Andreas **Naumann (**1744–1826): Farmer and amateur naturalist. Father of Johann Friedrich Naumann.

Johann Friedrich **Naumann** (1780–1857): Leading ornithologist of German nineteenth century, also bird painter. **See Chapter Seven.**

Charles **Nègre** (1820–1880): Pioneer of photography, both technically and in aesthetics.

Friedrich Wilhelm **Nietzsche** (1844–1900): German philosopher and cultural critic. His reputation suffered from his popularity with the Nazis.

Christian Ludwig **Nitzsch** (1782–1837): professor of zoology at the University of Halle, with a particular speciality in the distribution of feathers.

Ernst **Nolde** (1867–1956): German painter and printmaker, member of the Expressionist group *Die Brücke*. **See Chapter Ten.**

Carl Joseph **Oehme** (1752–83): Physician in Leipzig. Translator of Buffon.

Lorenz **Oken** (1779–1851): German naturalist and ornithologist, influenced by Kant and Schelling and a leading proponent of *Naturphilosophie*. **See Chapter Six.**

Jean-Baptiste **Oudry** (1786–1855): French Rococo painter and engraver. **See Chapter Four.**

Richard **Owen** (1804–92): English biologist, palaeontologist and anatomist. Responsible for founding of Natural History Museum in London. Controversy with Darwin.

William **Paley** (1743–1805): English clergyman attempting to prove the existence of God through nature.

Peter Simon **Pallas** (1741–1811): Prussian botanist and zoologist holding a professorship in St Petersburg.

Thomas **Pennant** (1726–98): Welsh naturalist and zoologist, wrote and illustrated accounts of his scientific travels.

Charles **Perrault** (1628–1703): French author responsible for some of Europe's best known fairy-stories.

Claude **Perrault** (1613–88): French architect, physician and anatomist. Involved with Le Brun in the design of the east façade of the Louvre.

Pliny (23–79): Roman author and general, author of the celebrated *Natural History*; famously died off Pompeii during the explosion of Vesuvius.

John **Ray** (1627–1705): Parson-naturalist, later collaborator with Willughby. His career suffered from the imposition of church orthodoxy.

Oscar **Rejlander** (1813–75): Swedish photographer working in London. Worked with Darwin on *The Expression of Emotions*. Did motion studies of horses.

Wilhelm Heinrich **Riehl** (1823–97): German ethnographer and professor of 'Volkskunde' in Munich. Subsequently regarded as being unacceptably right-wing.

Rainer Maria **Rilke** (1875–1926): Major Austrian poet and prose-writer.

Henri **Rousseau** (1844–1910): French artist, associated with *les Fauves*.

Jean-Jacques **Rousseau** (1712–78): Swiss social philosopher, major figure of Enlightenment. *Émile* (1762) explores issues around the education of 'natural man'.

Eduard **Rüppel** (1794–1884): Naturalist and explorer based in Darmstadt; published a number of travel journals from Africa.

Étienne Geoffroy de **St.-Hilaire** (1772–1844): French naturalist and colleague of Lamarck. Associated with Oken. Member of Napoleon's scientific expedition to Egypt.

Friedrich Wilhelm Joseph **Schelling** (1775–1854): German philosopher, often overshadowed by Hegel, but exercising a major influence on European Romanticism.

Friedrich **Schiller** (1759–1805): Poet, dramatist and historian, closely associated with Goethe in Weimar.

Hermann **Schlegel** (1804–84): Ornithologist and director of Leiden Natural History Museum. Extensive travels in Asia.

Martin (Meister) **Schongauer** (c. 1450–91): Renowned Alsatian engraver and painter.

Carl Ludwig **Seeger** (1808–66): German landscape and genre painter.

Johann Michael **Seligmann** (1720–62): Painter and engraver, specialised in birds. Produced editions of Edwards, Catesby and Gesner.

Johann Georg **Siegesbeck** (1786–1855): Doctor and botanist, working in St Petersburg for the Russian Academy of Sciences. Known for opposition to Linné.

Josef **Smit** (1836–1929): Dutch artist and illustrator, worked for Schlegel and in London for Gould and with Wolf and others.

Frans **Snyders** (1579–1657): Flemish artist, with wide range of patrons and subjects. Known as the first specialist animalier.

Johan August **Strindberg** (1849–1912): Swedish dramatist and novelist, who experimented with natural science and was close to European Naturalism.

Johann Conrad **Susemihl** (1767–1846): German engraver and natural history artist. His major work was: *Teutsche Ornithologie oder Naturgeschichte aller Vögel Teutschlands in naturgetreuen Abbildungen und Beschreibungen* (1800–1817); also illustrated Oken's *Allgemeine Naturgeschichte* (1839f).

Erwin **Stresemann** (1889–1972): professor of Zoology at the Humboldt University in Berlin, also from 1920 director of the Zoological Museum in the city.

William **Swainson** (1789–1855): English ornithologist, entomologist and systematiser. Wide travels, settled in New Zealand. **See Chapter Six.**

Johanna Dorothea **Sysang** (1729–91): German copperplate engraver. Bird portraits among her (largely unexplored) work.

Coenraad Jacob **Temminck** (1778–1858): Dutch zoologist and museum director.

Johann Heinrich Wilhelm **Tischbein** (1751–1829): German painter, resident in Rome and Italy for more than twenty years.

Ferdinand **Tönnies** (1855–1936): One of the founding figures of German sociology. Unusual in sociology circles in coming froma rural background.

Jakob von **Uexküll** (1864–1944): Major biologist, both practical and theoretical. Best known for his concept of the *Umwelt* (environment).

Rudolf **Virchow** (1821–1902): Physician and pathologist, renowned for work on the cell. Strongly empirical and opposed to Häckel's over-enthusiastic propaganda for Darwin.

Carl **Vogt** (1817–95): Medical professor, radical politician in Frankfurt Parliament of 1848 (controversy with Marx). Supporter of Darwin.

Alfred Russel **Wallace** (1823–1913): Exploring biologist and naturalist, celebrated for his work in the Malay Archipelago, and for nearly beating Darwin to the theory of evolution.

Jan **Weenix** (c. 1640–1719): Dutch painter famous for his pictures of dead game.

Francis **Willughby** (1735–72): Ornithologist and mathematician. Radical understanding of impartial science

Johann Joachim **Winckelmann** (1717–68): celebrated archaeologist and art-historian, famous for his studies of classical art.

Josef **Wolf** (1820–99): German nature painter, moved to London in 1848 and established himself as a freelance painter. ***See Chapter Eight.***

Émile **Zola** (1840–1902): Leading French novelist and figurehead of Naturalism.

J.B. **Zwecker** (1814–76): German popular illustrator.

Index

Agassiz, Louis 97, 126, 131, 193
Albin, Eleazar 43, 127, 193
Aldrovandi, Ulisse 43, 193
Alpers, Svetlana 150
Animal character 48, 56, 58
   59, 65, 80, 108, 132
  characteristic poses 135
Animal chartar
  characteristic poses 42, 50
  86, 114, 117
Anker, Jean 12, 146
Anschütz, Ottmar 143, 153, 193
Anthropocentrism 60, 81
  136, 158, 176
Aristoteles 170
Arnold, Joseph 49, 193
Audubon, Jean-Jacques 21, 57, 90
  118, 138, 165, 193
  Buffon 13, 32, 90
  Oudry 56, 57, 64

Baldner, Leonhard 12, 21
  32, 36, 41, 55, 112, 115, 194
Balzac, Honoré de 59, 114, 194
Barraband, Jacques 64, 194
Bartlett, Abraham Lee 130

Batsch, August 94, 194
Baudelaire, Charles 145, 188
Bayfield, Gabriel 48
Beaton, Isabella Mary 37
Bechstein, Johann 46, 112, 194
  Buffon 68
  Classification 94
Bell, Charles 27, 194
Belon, Pierre 32, 34, 40, 50, 181
Benjamin, Walter 38, 144
Benn, Gottfried 160, 162, 194
Berger, John 144
Bewick, Thomas 27, 155, 194
Bible, the 18, 20, 23, 24, 31, 35
Birkhead, Tim 12
Blumenbach, Johann 21, 36
  44, 94, 194
Bock, J.C 119
Boel, Pieter 63, 195
Botanical drawing 28
Bourdieu, Pierre 144
Brady, Matthew 188
Brecht, Bertolt 24
Bredekamp, Horst 17
Brehm, Alfred 108, 125, 195
  &Darwin 108
  Darwin 132
Brehm, Christian Ludwig 98, 99

	104, 112, 195	Crary, Jonathan	144
Brisson, Mathurin-Jacques	71, 195	Cuvier, Georges	97, 104, 110, 195
Büchner, Georg	97, 126		
Buffon, Georges-Louis		Darwin, Charles	18, 34, 46
Leclerc de	34, 61, 67		85, 88, 104, 112, 153, 157, 168
	90, 93, 111, 115, 195	Expression of Emotions	140
Bird song	84	Illustration	130
bird-song	180	Illustrations	25, 47, 109
classification	13, 53	Jacobsen	19, 114, 160
Classification	69, 92	The Expression of Emotions	130
English translation	72	Wolf	62, 130, 135
German translation	70	Darwin, Erasmus	195
illustrations	73	Daubenton, Edmé-Louis	73, 196
observation	15, 22	David, Jacques Louis	65, 196
Observations	83, 110	Deiker, Carl F	154
personal style	58	Derby, Lord	127
Burton, Robert		Diorama	126, 153
Anatomy of Melancholy	34	Dollmer	95
		Dugmore, A. Radclyffe	141
			148, 187
Camper, Petrus	115, 195	Little Tern Plate	147
Catesby, Mark	95, 195	Dürer, Albrecht	13, 30, 55
Chain of being, the great	69		136, 151, 159
characteristic charater		A litle Owl	54
characteristic poses	37	A little Owl	21
Chodowiecki, Daniel	76, 195	Comparision with Frisch	136
Christian Ludwig II	56	Comparison with Frisch	50
Clark, Kenneth	39	Melancholia	34
Classification	13, 21, 43	Rhinoceros	48
	53, 59, 72, 82, 92, 111, 115	symbolism	14
Taxonomy	19, 93	Symbolism	25
Colour	26, 89	Dutch painting	12, 38, 57, 150
Colour photography	143, 146, 187	Flowers	69, 152
Colouring	22, 44, 47	Game Piece	35
	52, 74, 109, 128		
Cooking recipies	35		
Copley, John Singleton	65, 195		

Eberlein, Christian Eberhard	96, 124	Gabler, Ambrosius	119, 125, 132
		Plates	120, 121
Edwards, George	47, 196	Galton, Francis	93, 197
Eliot, George	130	Gartenlaube, Die	110
Darwin	35, 62, 131	Gauguin, Paul	165
Eliot, Thomas Stearns	168	Gerstäcker, Friedrich	165
Elliot, Daniel Giraud	124, 196	Gesner, Conrad	32, 43, 48, 62, 112
Emblems	32, 34	Gessner, Salomon	59, 117, 143
Emerson, P.H	144	Goethe, J.W. von	17, 19, 76, 109, 119, 125, 197
Ethnology	113, 115	Gogh, Vincent van	152, 161, 165
Ethology	18, 81	Gould, Elizabeth	48
Expressionism	151, 161, 166	Gould, John	48, 93, 155, 197
Eyck, Jan van	27, 196	Wolf	124, 128, 171
		Gray, G.R	127
Fabritius, Carel	12, 196	Gros, Antoine-Jean:	149
Falconry	28, 38, 57, 90, 155		
Fechner, Gustav	60	Häckel, Ernst	67, 105, 142, 197
Film	61, 143	Hamsun, Knut	156
Flower symbolism	32	Heade, Martin	164, 197
Forster, Georg	70, 196	Heine, Heinrich	100, 125, 197
Foucault, Michel	59	Herder, Johann Gottfried	94, 197
Fox Talbot, William Henry	140	Hergenroeder, J.M	119
Freud, Sigmund	166	Hiroshige:	165
Friedrich II, Holy Roman Emperor	24, 35, 95	Hoffmann, E.T.A	76
Frisch, Ferdinand Helfreich	47, 54, 151, 155, 197	Hofmannsthal, Hugo von	156, 198
		Houdecouter, Melchior	198
The Brown-Black Eagle Plate	88	Hübl, Freiherr von	143, 148
The Smallest Owl Plate	50	Humboldt, Alexander von	12, 19, 53, 70, 117, 119, 125, 145, 151
Frisch, Johann Leonhard	13, 41, 58, 59, 64, 73, 84, 87, 94, 112, 117, 146, 151, 163, 197	Hunting and its illustration	25, 35, 57, 77, 111, 129, 175
Frisch, Jost Leopold	42	Hybridization	45
Frish, Ferdinand Helfreich			
Dürer	22	Ibsen, Hendrik	154
Fromentin, Eugène	150	Identification	24, 43, 52

                65, 86, 87, 93, 104
                112, 115, 128, 136
Illiger, Johann   53, 112, 149, 198
Impressionism   151, 157, 159, 165
Ingrès, Jean-A.D   141, 198
Ischtar Gate   58

Jackson, Christine   12
Jacobsen   161, 166
Jacobsen, Jens Peter   19, 105
                           152, 198
   Darwinism   19, 114, 160
Jacopo di Barbari   27, 194
Japanese Art   165
Jizz   37, 52
Jonsson, Lars   12, 152, 198

Kafka, Franz   34
Kaup, J.J   127, 198
Keulemans, John Gerard   127, 198
Klein, Jacob   49, 71, 94, 199

La Fontaine, Jean de 61, 62, 81, 135
Lamarck, Jean-Baptiste 67, 131, 199
Lambourne, Maureen   12
Lamm, Marcus zum   24, 199
Landseer, Edward 23, 134, 136, 199
Lavater, Johann Caspar   62, 199
Le Brun, Charles   61, 81, 199
Lear, Edward   48, 54, 59, 155
                  163, 175, 185, 189
Leonardo da Vinci   27
Levaillant, François   164, 199
Liebermann, Max   152, 156
Liliental, Otto von   143
Liljefors, Bruno   14, 114, 129

                137, 150, 159, 199
  Eagle Owl   156
  Plate   157
  Science   159, 163
Livingstone, David   124, 128
Lodge, George   137, 188
Lombroso, Cesare: 140
Ludolff, Michael   49, 200
Ludwig, Heidrun   136
Lumping and Splitting   93, 100

MacGillivray, William   127
                               141, 200
Mach, Ernst   159
MacLeay, Alexander   101
Marc, Franz   165
Marey, Étienne   143, 153, 200
Martinet, François-Nicolas 73, 200
Martini, F. H.W   70, 74, 94, 200
Martini, Friedrich H.W   177
Meerwarth, Hermann 140, 149, 153
Melville, Hermann   165
Menagerie in Versailles   49, 55
                        57, 58, 71, 128
Menzel, Adolf   79, 200
Merian, Sibylla   200
Merrem, Blasius 24, 80, 94, 95, 200
Meyer, Bernhard &
Johannes Wolf   100, 119, 122
Möbius, Karl   113
Mullarney, Killian   11, 14, 201
Munch, Edvard   156, 201

Naturphilosophie   45
Naumann, Johann Andreas 77, 201
Naumann, Johann Friedrich   14

	21, 52, 77, 95, 100, 108, 122	Photography	14, 38, 117
	124, 126, 130, 147, 151, 159, 201		137, 139, 148
Plate	137	& colour	38
Nègre, Charles	141, 201	colour	143
Nietzsche, Friedrich	101, 152	Phrenology & physiognomy	19
	154, 159, 201		62, 115
Nissen, Claus	17, 137	Piper, Richard	150
Nitzsch, Christian Ludwig	109, 201	Playing cards	28
Nolde, Emil	150, 152	Pliny the Elder	20, 42, 202
	163, 168, 201	Poe, Edgar Allan	141
Plate	167	Pointillism	144
Nomenclature	72, 93	Popularization of science	95
Nyhart, Lynn	113, 126	Portraits	15, 23, 39, 135
Observation	15, 18, 45, 46	Quinary system	100
	54, 68, 72, 78, 80, 105, 110, 112		
Oehme, Carl	70, 75, 83, 94, 201	Ray, John	12, 42, 202
Oken, Loren	92, 126	Redouté, Pierre-Joseph	64
Classification	161	Rejlander, Oscar	143, 202
Oken, Lorenz	13, 69, 76, 92	Rembrandt van Ryn	13, 25
	100, 101, 109, 126, 127		30, 40, 55, 87
Classification	161	Plate	31
Oudry, Jean-Baptiste	55, 56, 66	Riehl, Wilhelm Heinrich	130, 202
	68, 81, 85, 90, 106	Rilke, Rainer Maria	156, 160, 202
	129, 155, 175, 201	Röting, Lazarus	28
Character of birds	87	Rousseau, Henri	167, 202
Plate	56	Rousseau, Jean-Jacques	17
Plates	74		77, 203
Owen, Richard	97, 202	Rüppell, Eduard	127, 203
		Ruskin, John	116
Paley, William	46, 73, 202		
Pallas, Peter	95, 202	Schama, Simon	27, 39
Pennant, Thomas	72, 202	Schelling, Friedrich	69, 203
Periodic Table	102, 103	Schiller, Friedrich	17, 203
Perrault, Charles	58, 175, 202	Schinz, Heinrich Rudolf	95, 100
Perrault, Claude	49, 63, 202		110, 183

Schlegel, Hermann 25, 87
 118, 127, 130, 134, 146, 159, 203
  Scientific Illustration 16
Schongauer, Meister 26, 27, 203
Schulz, Georg E.F 144
Science as a political threat 68
Seeger, Carl Ludwig 129, 203
Seligmann, Michael 47, 75
 124, 203
  Plate 89
Sèvres porcelain factory 63
Siegesback, Johann 69, 203
Smit, Joseph: 127, 203
Snyders, Frans 39, 154, 204
Specialization 18
Species, the idea of 17, 43, 45
 53, 83, 86, 89, 93, 95
 100, 104, 114, 117, 131
St.-Hilaire, Geoffroy de 97, 203
Stresemann, Erwin 12, 204
Strindberg, August 154, 204
Susemihl, Johann C 124, 204
Susemihl, Johann Conrad 98
Swainson, William 48, 100
 102, 104, 107, 116, 131, 204
  Oken 94, 101
Symbolism 26, 32, 54, 159, 168

Sysang, Johanna 204

Taxonomy 19
 classification 21
Taxonomy see classification 60
Temminck, Conraad 64, 105, 199
The Brown-Black Eagle Plate
 Dürer 50
Tischbein, Johannes H.W 155, 204
Tönnies, Ferdinand 113, 204
Turnbull, Archibald 137
Typical, as a category 54, 63
 115, 116

Uexküll, Jakob von 133, 204

Virchow, Rudolf 67, 98, 131, 204
Vogt, Carl 68, 126, 205
Voss, Julia 23

Wallace, Alfred R. 157, 205
Wallace, Alfred Russel 124
Weenix, Jan 57, 205
Weiditz, Hans 28
Willughby, Francis 12, 42, 205
Winckelmann, Johann 155, 205

Zwecker, J.B 155